FIRE! ITS MANY FACES AND MOODS

The cry of Fire! startles everyone into being alert and taking action. Fire can cause disasters, but we depend on it in one way or another every day of our lives. Whether it is for cooking, heating, making glass, or the steel frame of your bicycle, we need fire. *FIRE! ITS MANY FACES AND MOODS* tells you about many of the magical, practical, and tragic aspects of fire throughout history, from the time of the caveman to the age of nuclear energy. You'll read about how it has caused terrible disasters like the great Chicago Fire, as well as learn about the numerous ways in which fire works for you.

BOOKS BY JAMES J. O'DONNELL

EVERY VOTE COUNTS
A Teen-Age Guide to the Electoral Process

FIRE!
Its Many Faces and Moods

GOLD
The Noble Metal

FIRE!

ITS MANY FACES AND MOODS

JAMES J. O'DONNELL

Julian Messner   New York

Copyright © 1980 by James J. O'Donnell

All rights reserved
including the right of reproduction
in whole or in part in any form
Published by Julian Messner
A Simon & Schuster Division of
Gulf & Western Corporation
Simon & Schuster Building
1230 Avenue of the Americas
New York, New York 10020

Julian Messner and colophon are trademarks
of Simon & Schuster registered in the U.S. Patent
and Trademark Office.

Manufactured in the United States of America

Library of Congress Cataloging in Publication Data

O'Donnell, James J
  Fire! its many faces and moods.

  Bibliography: p. 183
  Includes index.
  SUMMARY: A survey of the many aspects of fire including its uses, magical associations, destructive potential, prevention, and how we depend on it every day.
    1. Fire—Juvenile literature.   2. Fire extinction—Juvenile literature.   [1. Fire.   2. Fire extinction]   I. Title.
QD516.0217   541.3'61   80-10123
ISBN 0-671-33021-7

for
Kristen, Lisa, and Elizabeth

CONTENTS

	Introduction	9
1.	The Gods of Fire	13
2.	The Magic of Fire	24
3.	Putting Fire to Work	33
4.	Fire Goes to War	43
5.	Fire Disasters	54
6.	Volcano	67
7.	Forest Fire	80
8.	Crime Without Pity	87
9.	Fighting Fire	96
10.	Fire Fighters	111
11.	What is Fire?	120
12.	The Personality of Fire	126
13.	The Mighty Match	140
14.	The Agony of Fire	148
15.	The Constant Menace	153
16.	Home Inferno	163
	Epilogue	175
	Glossary	178
	Suggested Further Readings	183
	Index	185

INTRODUCTION

Of all the things that make up the earth, which do you think are most vital for the survival of the human race?

Since the earliest of times, this question has occupied the thoughts of the philosophers, men and women whose main joy in life is the discovery of knowledge and the contemplation of truth.

If you were asked to rank the four most important things, what would your choices be? The task is more challenging than it first appears to be.

At the very top of the list you would have to rank air. Without air, breathing would be impossible, and the human race would suffocate. Few of us can hold our breath for even a minute without gulping for life-sustaining oxygen. We can be grateful for the thin mantle of atmosphere that is bound to the earth by the pull of gravity.

The land itself ranks among the most basic needs of humankind. Going out to sea might be fun, but only if there is a port where you can set your feet. It is easy to take the land for granted, but some scientists are seriously concerned that some day we might

FIRE! ITS MANY FACES AND MOODS

lose large chunks of the continents that we inhabit. Changing climate patterns could melt polar ice to the point where half the cities of the world would lie beneath the sea.

Even though the human race is threatened by water, we nonetheless must list water as one of the most important of earthly commodities. The human body is 90 percent water, and whereas we can go for days without food, lack of water for even one day can cause dangerous dehydration of the body. Vegetation won't grow without water, and without grasslands and cultivated crops no living creature would survive.

Since ancient times philosophers have listed air, earth, and water as three of the four fundamental elements. You might be surprised to learn what they considered to be the fourth. Actually it is more of a force than a substance. It is fire.

Everyone is fascinated by fire. Who can ignore the beauty or escape the hypnosis of a fire burning in a fireplace? You can be sure to find a crowd gathered when you hear fire engines' wail. Chances are, however, that you have never given too much thought to the nature of fire or its importance in the framework of life. It is something you may take for granted, recognizing that it can serve both as a comforting friend or as a fearsome foe. It might sound strange, then, to be told that there is virtually no aspect of your life that is not influenced in some way by fire.

Introduction

Two obvious uses of fire are for heating your home and cooking your food. But there were vast stretches of time when it never occurred to anyone that the power of fire could be tamed. People knew how to warm themselves only by wearing the skins of animals, and they ate uncooked meat. During those generations fire was an enemy, a mighty god of destruction and terror from which all had to flee in order not to perish.

Even today in some regions of the world, men, women, and children live in mud huts simply because they have not learned that mud can be fired inside kilns and turned into bricks. Most wooden houses are held together by metal nails made possible only because of the roaring fires of blast furnaces in our industrial cities.

Your family car is run by the force of fire, and this is true also of airplanes, locomotives, ships, and spacecraft. And don't kid yourself by thinking that electricity can be used to replace the energy of fire, because most electricity is generated through the power of fire.

Fire is more than important. It is exciting and mysterious. The more you think about it, the more rapidly questions arise. Just what is fire? How is it produced? How is it that we blow into a fireplace to stimulate a fire, whereas we blow on a match to put it out? Are explosions a form of fire? Why is it that some substances burn and others won't? Is it true that there is

FIRE! ITS MANY FACES AND MOODS

a material that must be stored underwater because it bursts into flames when exposed to air?

Fire has played a major role in the long history of the human race. It goes back to the time when each wandering tribe had a member whose sole job was to keep the fire alive, death being the penalty should it go out. Fire is closely related to religion; it is not by chance that candles are still used in religious ceremonies. Uncontrolled fires have demolished most of the world's major cities at one time or another and have caused countless other flaming disasters. And from the day that a clever archer conceived the idea of a flaming arrow, warfare has never been the same. Is it any wonder that the philosophers listed fire as one of the four elements?

Indeed, the story of fire could fill a book. And you have that book in your hand.

1.
THE GODS OF FIRE

Imagine for a moment that you are a citizen of Rome in the days of Julius Caesar. You have just returned from a trip abroad to the Roman colony of Gaul, the land now known as France. You have been away for a long time, and now at last you stand again on the soil of your homeland.

What is the first thing that you will do? Run to meet your parents, brothers, and sisters? Look for old friends? Find the one who promised to wait for you? Perhaps you'll do all of these things in due time, but not until you fulfill the most important of all obligations: you must kneel before the sacred flame and give thanks for your safe return!

In ancient Greece and Rome, every home had an altar upon which a flame glowed all the time. It wasn't a mere symbol of worship; the flame was the actual living god of the household. It could be fed only with the purest of fuel, and no unworthy act could ever be performed in its presence. It was allowed to go out only if the entire family had perished.

You wouldn't be allowed to pray before the sacred fire burning in the home of a friend. Unlike the present when multitudes of people worship a single god, the Romans believed that each family had its own personal god. The flame was thought to contain the surviving spirits of the family's departed ancestors. Prayer was private, and rituals were kept secret from anyone who was not a member of the household.

The nature of such a religion made marriage a very serious affair. From her earliest childhood a young girl worshipped only at the altar of her father. When the time came to marry, it meant that she had to give up the god of her parents and embrace a new god, the god of her husband's family.

The most solemn part of the marriage ceremony didn't take place in the temple, but rather in the home of the groom. The bride knelt before the flaming god of the hearth and prayed to be accepted. Some contemporary marriage ceremonies show many traces of Roman origin.

Candles still burn today in many churches, the perpetual flame being a symbol of an everlasting god. People who kneel in church to pray often light votive candles, a way of keeping their petitions alive.

Because the licks of flame contained the spirits of the ancestors of a Roman family, it was a grave sin to let the fire go out. In effect, such negligence would

The Gods of Fire

be the mass murder of all the spirits of the past. For the same reason it was a serious sin not to have children. They were needed to keep the fire alive in the future.

This logic made marriage an obligation. Love and happiness were of secondary importance. The primary purpose of the union was to perpetuate the family and keep the sacred flame kindled. After the birth of a child a few days were allowed to pass, and then the entire family assembled before the altar flame to present the child to his or her forebears.

The Roman concept of a flaming god for each family extended to the devotion to a universal flame that protected entire cities. In Rome the goddess of fire was Vesta, and a temple dedicated to her stood in the center of the city. It was attended to constantly by six *vestal virgins*, young girls whose lives of dedication were highly honored. If a Roman family moved to a different city, they would go to the temple and light a flame that they would carry with them to their new home. In this way they could continue to worship the same god.

Perhaps it is difficult for you to understand how people could believe that a flickering flame was a living god. You, however, are living in an age of science and understanding, whereas the most knowledgable of the ancients were less informed factually than schoolchildren of today.

Also, think for a moment about the dance of a flame.

FIRE! ITS MANY FACES AND MOODS

Where else in nature is there a more subtle display of living movement? Flames reach for the heavens, and crackling embers speak a secret language. Fire gratefully consumes every sacrifice offered to it, giving thanks by extending its arms and increasing in brightness. Fire gives comfort as a god should. Haven't you ever been hypnotized by the glow of a fireplace in a darkened room or by a campfire beneath a star-filled sky? Finally, fire has the quality of eternity. The eternal flame that burns at the grave of our assassinated President Kennedy is a living proclamation that his spirit survives.

In the early years of history there was practically no scientific understanding. Mysteries of nature had to be explained with poetry and mythology. How do you think, for example, the Greeks explained the origin of fire?

Prometheus, according to the myth, was a high-ranking god. He and his brother, Epimetheus, were assigned by the other gods to distribute to different animals all the powers they would need in order to survive. Epimetheus did most of the work, but he was a bit too generous. By the time the human animals' turn came, there were no gifts left.

Prometheus felt sorry about this, so he made a bold decision. He stole fire from the gods and presented it to humankind as their gift.

When Zeus, the highest of all Greek gods, heard

The Gods of Fire

what Prometheus had done, he became very angry. As a punishment, Prometheus was chained to a boulder on top of a mountain. Each day an eagle swooped down, tore out Prometheus' liver, and consumed it. Each night the liver grew back again. In this fashion Prometheus suffered for thousands of years. He was finally rescued by Hercules, the most famous hero of Greek legend.

The early Greeks weren't alone in inventing tales as to how humans got the gift of fire. Primitive tribes in every corner of the earth made up their own stories, and they are remarkably similar. Usually the legends are fantasies of bravery in which a hero soars skyward on a magic bird to steal fire from the sun and bring it back to earth.

In truth, the discovery was probably nowhere near as spectacular. The role of Prometheus was most likely played by a bolt of lightning.

Greek philosophers spent a lot of time trying to figure out the ways that things work in nature. Why, for example, is pepper hot and honey sweet? Their answer to this question was an amazing anticipation of secrets to be discovered thousands of years later. The philosophers reasoned that all matter is made of tiny atoms. Pepper atoms have sharp edges, whereas honey atoms are smooth. The speculation was rough around the edges, but it was a start.

Heraclitus, like many other philosophers, was

FIRE! ITS MANY FACES AND MOODS

troubled with a nagging question. How is it that things are constantly changing, yet they seem always to remain the same. For example, it is impossible to swim twice in the same river. The river might look the same as it did yesterday, but the flow of the current has made it entirely new.

At first glance the problem might seem to be little more than a riddle, but the insight is profound. From modern science we know that even our bodies are completely replaced every seven years or so because dying cells are constantly being replaced by new ones.

Because stability exists in the midst of change, Heraclitus felt that all things must be made up of a basic element possessing conflicting characteristics. It was obvious to him that this element must be fire.

After all, look at a burning candle. You would swear that the flame is constant, and that you can look at the same flame for minutes on end. Look closer, however, and you will see that the flame is in a constant state of strife and change. In no two instants is it actually the same. In varying degrees isn't this the same with everything? The rusting of a nail, except for tempo, is no different from the holocaust of a forest fire!

It wasn't until two thousand years after the death of Heraclitus that experimental science came into being. It started when the alchemists of the Middle Ages attempted to change base metals such as lead into gold. Additional centuries were required, however,

The Gods of Fire

before humankind freed itself of the shackles of magic and superstition.

If you were sent back to live in the Middle Ages, you'd find life a lot different. You wouldn't have movies, radio, stereo, television, or telephone. Major-league baseball and super-bowl football didn't exist. In such a setting, how would you find entertainment?

Possibly some of the most exciting events would be the fire festivals that were held at special times during the year, particularly at the beginning of the planting and harvesting seasons. If you wonder whether you'd prefer a fire festival to a football game, here's an outline of a day a seventeenth century teenager would have considered exciting.

It is the first Sunday of Lent, the forty-day season before Easter Sunday on the Christian calendar. For the past week you and your friends have been going from house to house and into the fields to collect straw and brushwood. Each day you've carried the collection to the top of a nearby hill, piling it up around a tall, slim birch tree. You've nailed a plank of wood to the tree to make it in the form of a cross. Now it is the day of the fire, and excited residents of the entire village have gathered on the slopes of the hill. The fire is set, and soon flames are leaping high into the sky. You and your friends light torches and march around the fire, praying aloud. Everyone watches the column of smoke, because if it drifts in the direction of the wheat fields, it is a sign from heaven that there will be a bountiful crop. When the fire gets lower, you

FIRE! ITS MANY FACES AND MOODS

jump over it for good luck. If your family owns a cow, you lead her through the glowing embers to insure her health and fertility.

Sound exciting? Or are you glad that you live in the twentieth century?

Easter Sunday calls for another fire on top of the hill. All the fires in the village are extinguished, and one by one the villagers go to the hill to bring the new flame home to their hearth. The Easter fire will protect the home against fire, lightning, hail, and disease throughout the year to come. Charred sticks are planted in the fields to protect the crops against rodents. Ashes are saved to be mixed with the seeds when planting time comes.

Vestiges of these customs survive today in the ritual of the Christian church. At Easter time the Paschal Candle is lighted to start a new church year, and the palms that decorate the church on Palm Sunday are saved for the next Lenten season when they are burned to ashes on Ash Wednesday. These ashes are smudged on the foreheads of the faithful to remind them that some day they will return to dust.

The traditional burning of the yule log during the Christmas season is also a carry-over from the fire rituals of the Middle Ages. Because winter is cold the logs were burned indoors, making the event a family affair, and this is the way the custom survives today. It's not too likely today, however, that families pray

The Gods of Fire

aloud for protection against lightning, mildew, fire, vermin, witchcraft, blight, and infertility.

If fire has often played the role of a god, it is equally true that it has often come closer to resembling the devil. At one time the burning of living humans to placate the gods was a common practice.

The Roman conqueror Julius Caesar wrote about the human sacrifices practiced by the Celts in Gaul. Condemned criminals were set aside to be sacrificed to the gods at a great festival that was celebrated every five years. Prisoners of war were included because blessings increased according to the number of victims. A huge wicker-work structure of wood and grass was built, and on the day of sacrifice the captives were herded inside and the doors were sealed. Then the fire was set. Screams of agony filled the sky, but the joyous audience felt no pang of sympathy or guilt. The greater the suffering, the greater would be the fertility of the land.

When the Spanish conquistadors invaded the New World following the discovery of Columbus, they sent back to Spain tales of human sacrifice practiced by the natives. Revolted by the pagan rituals of death, Spaniards failed to feel guilty themselves concerning the horrendous era of the Inquisition. Under the guise of rooting out religious heresy, the Inquisition put thousands to death by *auto-da-fé*. The words mean "act of faith," but it could be described better as an

FIRE! ITS MANY FACES AND MOODS

act of satan. It was death by public burning. The victim was tied fully conscious to a wooden post, and kindling was piled high at the base of the stake. The fire was then set, and milling crowds cheered and reveled amid the cries of pain. During an eighteen-year period in the fifteenth century more than ten thousand persons suffered this horrible execution.

Auto-da-fé was not confined to Spain. You probably know the story of Joan of Arc who, guided by voices from heaven, led the French army to victory over the English. Her execution was carried out in France by her English captors in 1431, and thousands of persons in the marketplace of Rouen saw her die. The injustice of her fiery execution was acknowledged in 1456 when she was pronounced innocent by Pope Calixtus III. In 1920 Joan of Arc was canonized (made a saint) by the Catholic church.

Fire was also used during the Middle Ages as a test of innocence. A person suspected of a crime was required to carry burning coals bare-handed, lick a white-hot iron, or thrust an arm into a flaming fire. If no injury resulted, it was believed that God had spoken in the form of a miracle, and the suspect was declared innocent. Most often, of course, the test resulted in painful burns that were quickly relieved by death. There was no need for lengthy trials.

The ultimate relationship between religion and fire is embodied in the concept of hell. Some theologians

The Gods of Fire

question whether hell exists as a place of actual physical pain. Others insist that the pits of hell, described in the scriptures and in literature, really exist as the place of reckoning for those who have led sinful lives.

No one, while on earth, will ever be certain which view is correct. Only one thing is sure. If there is an eternal punishment that is physical, then it certainly exists in the form of fire, the most awful of human afflictions.

Since the beginning of time, man has looked at fire with wonder at the mystery and magic it inspires. Let's look next at some of that magic.

2.
THE MAGIC OF FIRE

In one of the temples of ancient Greece a heavy stone statue of a goddess stood high on a pedestal. She held a wine jug in her hands, and at her feet there was a place to kindle a fire as an expression of worship. It was common practice to pour small libations of wine into such a fire as an offering. This particular statue, however, was special. Shortly after a fire was lit, the goddess herself would pour wine from the pitcher she held onto the flames burning below. To everyone who witnessed this phenomenon, it was clearly a miracle.

As with all magic, however, there was a simple explanation. A small airtight compartment was built into the pedestal just under the fireplace, and this was kept filled with wine. When a fire was ignited the heat caused the wine to expand, forcing it upward through a tube leading to the mouth of the jug. All of these workings were hidden from view, and the appearance was that the wine flowed from the pitcher. A simple

The Magic of Fire

property of fire and heat was exploited to astound countless worshippers.

A similar device is explained in the writings of a Greek, named Hero, who lived more than two thousand years ago. This time a fire was lit at the entrance of a temple, causing water in a hidden reservoir to expand and flow through a tube to a bucket suspended by ropes. As the bucket filled it became heavy, pulling the ropes downward and activating a system of ropes and pulleys, again hidden from view, that swung open the temple doors. After worship the fire was extinguished, and the air in the reservoir contracted, drawing the water back out of the bucket and causing the doors to close. Once again, visitors to the temple considered the effect to be the result of divine intervention.

The use of fire to produce magical effects is probably as old as humankind. An account of fire-breathing goes back to 134 B.C. A Syrian named Eunus attempted to organize a slave uprising, and to demonstrate his power he astounded his followers by blowing jets of flame from his mouth. His trick depended on a hollowed walnut with a small hole at each end. He filled the shell with glowing embers and sulphur, secreted it inside his mouth, and by blowing through one end he caused fire to shoot out the other. The slaves were thrilled, but not the Roman consul, Publius Pubilius, who smothered Eunus along with the flames for good measure.

FIRE! ITS MANY FACES AND MOODS

Fire-eating was developed during the Middle Ages by traveling magicians and court jesters. They would take a pan filled with glowing chunks of charcoal, spear one with a fork, and put it into their mouths as though it was nothing more than a french fry. Then a spoonful of flaming oil would be taken from a bowl and swallowed just as casually. For dessert, molten metal would be poured on the tongue and chewed until it cooled and became hard.

Carnival performers today put on the same act. Sometimes they protect their mouths with alum, soap, or other ingredients, but the main trick is a simple deception. Pieces of white pine are mixed in with the burning charcoal. When blackened they look the same, but they hold little heat. Being a soft wood, the pine is easy to find with a fork, and if the mouth is kept well-moistened with saliva, the discomfort is not much greater than from eating something hot from a grill.

The drinking of the burning oil is pure illusion. When the spoon is removed from the flaming bowl, there is only a thin film of oil clinging to it. As the spoon enters the mouth, a quick breath blows out the flame before any damage can be done.

As for pouring molten metal onto the tongue, the result would indeed be awful if one of the more common metals were being used, but an alloy of bismuth, lead, and tin has a very low melting temperature. If the tongue is kept wet, the effect is no more uncomfortable than taking a sip of hot coffee.

The Magic of Fire

A French magician, Robert Houdin, reportedly could dip his hand into a cauldron of molten iron. By having his hand wet, a layer of insulation was formed by steam, just as a drop of water hops around on a hot skillet, protected by a cushion of steam.

If you are thinking of trying any of these tricks yourself, forget it immediately. Professional entertainers have years of apprenticeship and experience. When amateurs try to imitate the act, the result is often tragic. Recently, a member of a rock singing group breathed out jets of flame as part of the act. A young lad in the audience got the idea that he could do the same thing. When he arrived home he went into the garage, took a mouthful of gasoline, and lit a match. The explosion nearly killed him, and for the rest of his life he will suffer for not remembering the admonition he must often have heard as a child, "Never play with fire."

Magic illusions using fire are easy to see through if you concentrate on the nature of fire and its behavior. Many years ago a Frenchman named Julian Xavier Chabert became an international sensation by apparently defying the laws of nature. He would build a large iron furnace in which he lit a blazing fire. Then he would take a raw steak in each hand, enter the chamber, and close the door behind him. Several minutes later the door would open and Chabert would emerge, his body unburned and his clothes unscorched, but now holding two sizzling broiled steaks.

There seemed to be no possibility of deception.

FIRE! ITS MANY FACES AND MOODS

Close examination of the chest was allowed, and there were no secret compartments or trapdoors. A thermometer registered an inside temperature of 900° C, and water boils at 100° C. Still, the magician's secret was remarkably simple.

The firebox was located halfway up the inside of the oven. When Chabert entered the oven he placed the steaks directly over the flames and then curled himself on the floor of the furnace, covering himself with his heavy cloak. He breathed air through a small hole in the bottom of the chest.

Heat rises, and although the upper part of the oven reached a high temperature quickly, the lower portion remained relatively cool for a short time. In a few minutes the steaks were seared and it was time for Chabert to grab them, throw open the door, and astonish the audience.

The same principle is used to accomplish a feat of magic that originated in the Orient. A roaring fire is lit beneath a deep cauldron of water, and soon the surface is bubbling with steam. At that moment a fakir jumps into the tank and disappears below the surface.

You've seen lobsters cooked in this manner. How can you explain it then when a few minutes later the fakir leaps from the tub completely unharmed?

The explanation is that the trick works only when the tank is deep and the water is cold to start with. Again, heat rises, and when the water first boils on the surface, the water at the bottom is comparatively cool. When the performer jumps into the vat, the cool water

The Magic of Fire

mixes with the hot, and the overall temperature is little more than that of a comfortable bath. Do you believe it? You're better off not trying it out yourself!

Fire-walking dates back to antiquity, and today it is still practiced in the East, mostly as a tourist attraction. When you see a barefoot person run the length of a bed of glowing flames, your first thought is that it must be a trick. Either the flames aren't real, or else the feet must be covered with a fireproof material. Strangely, however, there is no such deception.

In the 1930s a group of English scientists conducted some experiments, inviting an Indian fire-walker to demonstrate his skill. A 10-foot trench was filled with wood and coal and set on fire. The fire was allowed to burn until the surface temperature was 430° C and the interior temperature was 1400° C. The fire-walker's feet were inspected, and although they were toughened from practice, there was no secret coating of any kind. The man then astounded the scientists by walking the length of the burning pit four times, and when he had finished his feet didn't show a single blister.

He didn't fare as well, however, when the length of the trench was doubled, and several blisters did form. This revealed to the investigators the secret of fire walking. It was a matter of exposure limit. If no more than two steps were taken with each foot, and if each step touched the fire for only a fraction of a second, toughened feet could stand a temperature as high as 2700° C. Exposure beyond that, however, would cause

FIRE! ITS MANY FACES AND MOODS

burns because of the cumulative effect of the heat.

Some entertainers succeed in conditioning their bodies to withstand heat and pain. An Italian physician and chemist named Sementini prepared his skin with a certain acid to the point where he could apply a red-hot iron to his skin with no harmful effect. Again, however, viewers could observe that the contact was only momentary. When a person irons clothes, it is common to test the heat of the iron by touching it for just an instant. The principle is the same.

Flame-eaters in medieval Europe were sometimes hanged for practicing witchcraft. The effect, however, is anything but supernatural: some liquids burn at a low temperature, and the trick is never to inhale. When the mouth is closed on a flaming torch, the supply of oxygen is cut off and the flame is extinguished immediately. Again, however, don't try it yourself. Even professionals suffer an occasional painful burn.

Fire figures prominently in common superstitions. If you have a fireplace in your living room or den, chances are that you live in a pretty modern home. A century ago it was common to have a fireplace in every room. Instead of watching television, which wasn't yet invented, people often sat watching the fire. Imaginative games and stories were created and superstitions flourished.

When a cinder leaped from the fire onto the floor,

The Magic of Fire

it was allowed to cool and then examined carefully. If it had the shape of a rectangle it represented a coffin, and someone was soon to die. An oval stood for a cradle, meaning that a baby was due to be born. A round cinder was a purse, and this meant that money was on the way.

If the fire was burning on only one side, a wedding would soon take place. A crackling fire predicted an early frost. And if soot fell from the chimney, a great disaster was in the offing. Beware of smoke coming from the chimney. This meant that a witch had taken up residence in the chimney, and the only way to chase her away was to form a cross by laying the poker across the bars of the grate.

Blowing out the candles on a birthday cake to make a wish come true is a ritual practiced by almost every family. Country folk in America used a wooden match to find out if their sweethearts were faithful. A flaming match was held upright, and if the flame curled toward him or her, everything was fine. In England if a girl felt unsure of her boyfriend Harry, she would insert two pins through the wick of a candle or else throw twelve pins into the fireplace at midnight, reciting the following as she did so:

> 'Tis not these pins I wish to burn,
> But Harry's heart I wish to turn.
> May he neither sleep nor rest,
> Till he has granted my request.

FIRE! ITS MANY FACES AND MOODS

The way candles burned also had a meaning. A candle burning on one side meant death, and a blue flame indicated the presence of a spirit in the room. It was very unlucky to look into a mirror by the light of a single candle. If you lit a second one, all would be well.

The invention of electric lighting has certainly made life a lot easier, but at the same time it has done away with much innocent fun.

We depend on fire for many practical uses. In the next chapter you will see some of the ways in which fire works for you every day.

3.
PUTTING FIRE TO WORK

Your ancestors learned to put fire to work several hundred thousand years ago. A roaring fire set just inside the entrance of a cave served several purposes. It cooked food, it provided warmth, and it gave off light. It also frightened away wild animals at night.

We still enjoy the first two benefits of fire. Of course, electricity provides us with much more efficient lighting than fire, and most of us don't have to guard against wild animals. New uses, however, have more than made up for these. We use fire to change clay into pottery and brick, to release metal from ore, to forge steel, to vulcanize rubber, to generate electricity, and to drive machines. The list could go on and on.

The single event that did most to give birth to the Industrial Revolution and the world of modern technology was the invention of the steam engine.

The popular tale that James Watt discovered the power of steam by observing the way that it lifted

FIRE! ITS MANY FACES AND MOODS

the cover of a tea kettle has one significant flaw. Watt was not the first to make the discovery. The first known steam engine was described by Hero in the year 120 B.C. It was a small hollow globe connected to a steam kettle. On opposite sides of the globe were two exhaust tubes pointed in opposite directions. As the steam escaped from these outlets, the jet action caused the globe to spin. Hero's steam engine was little more than a toy and served no practical use, but it was a start.

In the seventeenth century a jet vehicle was designed using the power of steam. Drawings show a four-wheel carriage upon which is mounted a firebox and a spherical boiler with an exhaust pipe pointing to the rear. Obeying Newton's third law of motion, which states that every action has an equal and opposite reaction, the force of the steam shooting backwards would push the vehicle forward. The same principle is used to fly today's jets, but the first proposed jet carriage never got off the drawing board and into production.

The primary energy that lies hidden in fire comes from the fact that fire boils water, boiling water produces steam, and steam creates pressure by expanding and contracting according to changes in temperature. This is what causes the lid on a pot of boiling water to jiggle. If the cover were sealed so that the steam had no escape, pressure would build inside the pot, causing it to explode with great force. The challenge of

Putting Fire to Work

science was to figure out a way to control this force and to put it to useful work.

In 1601 Giovanni della Porta put steam to work without creating the force of pressure. When a chamber of steam is cooled, it contracts as it condenses back into water. This creates a vacuum, and the resulting suction provides a pulling force rather than a pushing force.

This principle was used in the design of a pump invented in 1698 by an Englishman, Thomas Savery. At the time, England's coal mining production was being threatened by flooding in the mines, and Savery's pump came just in time to stave off economic disaster. The pump came to be called "The Miner's Friend."

Another Englishman, Thomas Newcomen, invented a more efficient steam-driven pump in 1712. The ends of a seesaw beam were supported by piston rods leading down into two cylinders. As steam pressure built in one cylinder, the expansion pushed up that end of the seesaw, lowering the other end. Cold water was then sprayed on the hot cylinder, making the steam condense and creating a vacuum, which pulled the piston down. This reversed the motion of the seesaw, and a sustained rocking kept the pump in operation.

It was while repairing a Newcomen engine that James Watt made his great discovery. He saw that steam was wasted each time the heated cylinder was cooled. After several experiments he came up with an engine in which the cylinder and the condenser were kept separate. The cylinder remained hot all the time, saving the tremendous amount of fuel previously needed to reheat the cylinder for each stroke of the engine. Watt's engine was eight times more efficient than the Newcomen engine, and, more important, it could be used to run any machine, not just pumps.

By converting the up-and-down motion of the piston rods to the circular motion of a flywheel, infinite possibilities were opened, because rotating wheels can do many types of work. It was only a matter of time before technology would bring about railroad locomotives, steamboats, steam fire engines, steam shovels, and many more. At one time the Stanley Steamer, a steam automobile, enjoyed great popularity. In our era of fuel shortages, perhaps the steamer will make a reappearance on American highways.

Over the years the efficiency of steam engines has increased dramatically because we have learned to use high pressure and superheated steam safely. Today a pressure of 1,000 pounds per square inch is common.

You might be tempted to say, "Who needs steam engines anymore? We live in an age of diesel trucks and jet airliners."

You'd be right—up to a point. In the field of transportation vehicles the internal-combustion engine has taken over. This doesn't mean that fire no longer plays a role, because the cylinders are moved up and down

Putting Fire to Work

by the internal explosions of fuel. In a gas engine the fuel is ignited by electrical discharges from the spark plugs. In a diesel engine the heat is supplied by the compression of the fuel vapors in the cylinders. Jet airplanes are propelled by the powerful discharge of superheated fuel inside the engines. But don't think for a moment that the power of steam has lost its importance. You probably rely upon it every time you switch on a light.

Electricity isn't something that is dug out of the ground. It has to be generated, which means that giant wheels must be made to turn. In a hydroelectric plant this can be done by using the force of moving water, such as that which is harnessed at Hoover Dam, which controls the flood waters of the Colorado River and provides electric power for a large area of the Pacific Southwest. If you and your family go camping, perhaps you generate electricity with a small gasoline-driven generator. Most of the electricity that supplies the towns and cities of America, however, is produced through the efficiency of steam turbines.

A turbine is nothing more than a wheel that is turned by a moving fluid—steam, air, and gases being included under the heading of fluids. For centuries, windmills and waterwheels have been turning wheels to grind wheat and run machinery. A waterwheel dips into a stream and turns at the same rate as the stream, changing the linear motion of the stream into the circular motion of the wheel's axle. Using gears or belts,

FIRE! ITS MANY FACES AND MOODS

a second wheel can be connected to the first, and if it is a smaller wheel it will turn faster. In this way a series of wheels or gears can be connected so that the final wheel will deliver either speed or power, whichever is needed.

Steam turbines are among the most powerful machines in the world. A single steam turbine connected to a generator can provide enough electricity for a city of three million people. High-pressure steam enters one end of the turbine and expands as it rushes through a series of meshed wheels. The wheels change in size until they are spinning faster than the speed of sound.

And how is the steam produced? Once again we come back to fire. Hydroelectric plants can be built only where huge dams can be constructed, and it will be a long time before nuclear plants can be made safe enough to take over the job now being done by burning coal, oil, and gas. If you ever visit a generating plant you will be fascinated by the mammoth turbines which dwarf their attendants. Perhaps you will be allowed to peep into the giant furnace, a raging sea of white-hot flame that is the raw material of electricity.

There is another fire at your constant service in the basement of your home. It is the furnace that provides the heat and hot water you take for granted. How does it work?

Some systems take hot air directly from the fire chamber and circulate it through a duct system into

Putting Fire to Work

the different rooms in your house. Each room has a hot air register built low into one of the walls. Other systems use the flames to heat a tank of water that circulates through a system of pipes and radiators. Not quite as common are systems that circulate steam through the pipes and radiators.

Let's take a quick look at a hot-water heating system. Once again, you start with fire. Years ago almost all furnaces were coal-fired with the coal fed by hand, shovelful by shovelful. The invention of the automatic stoker, a long, turning, screw device that carried coal into the firebox, eliminated a lot of hard work, but consistent burning was hard to achieve and ashes had to be collected.

Most systems now burn heating oil or natural gas. In either case, the fuel is sprayed into the firebox in the form of a fine mist and is ignited by an electric arc generated in the head of the burner. The flames heat water in coils of tubing which are connected to a water tank above the furnace. Hot water rises, and as it does it is replaced by cool water, which in turn is heated. When the water in the system reaches a set temperature, a thermostat turns off the system.

An electric-driven circulator carries hot water to the radiators in each room where the water gives off its heat and then returns to the tank to be reheated. Another thermostat inside the house is set to turn off the system when the rooms are comfortably warm. During the summer, hot water is not circulated through the

FIRE! ITS MANY FACES AND MOODS

radiator system, being needed only to supply the hot water taps in the kitchen and bathroom.

When it comes to cooking, it is only in recent years that families have the option of using methods other than that of their forefathers. Cooking over the flames of an open hearth ended a long time ago, but perhaps your grandparents can tell you of their youth when the first one up in the morning had to light a fire in the kitchen stove before the house would get warm and breakfast could be made.

The gas range revolutionized cooking and is still in wide use today. Electric ranges don't produce flames, but rather produce heat by passing an electrical current through metal coils having a high resistance.

Now we have entered an age of microwave and convection cooking. These involve oven cooking but not range-top cooking.

It is estimated that one out of every four American families will be using a microwave oven by the end of the 1980s. This is quite a prediction considering the fact that they were introduced only in the late 1960s.

Microwaves are a form of electronic radiation. The metal interior of the oven reflects the waves, and as the waves bounce back and forth in the oven they are absorbed by the food being cooked. The water molecules in the food vibrate, the vibrations produce heat, and the heat cooks the food. Cooking time is shortened because only the food is heated, not the oven or

Putting Fire to Work

the air around the food. Also, microwave ovens use less than one-third of the energy needed in a conventional electric oven.

Using a microwave oven you can roast a piece of meat on a paper plate. Paper doesn't absorb microwaves, so the plate won't burn. Use a regular roasting pan, however, and you're in trouble, because the metal deflects the microwaves away from the food.

Convection ovens work differently. In these units the inside air is heated, circulated, and reheated in a continuous cycle. The heated air is temperature-controlled, and it is efficient because it strikes all surfaces of the food simultaneously. The oven is lightweight, and with an outdoor outlet it can be used on the patio as well as in the kitchen. Some cooking experts say that the flavor of convection-cooked food is superior, and that these ovens will be favored in the future.

A final word about the lighting in your home. In medieval castles light was obtained from burning bundles of straw that had been dipped in pitch (a tar-like substance that burns). For many centuries candles provided light, and then came kerosene lanterns and gas jets. The bulb that burns in your reading lamp is the final step in a long process of development. Throughout this process, fire has been the workhorse. One riddle, however, still has to be explained. What about electricity that is produced by

FIRE! ITS MANY FACES AND MOODS

hydro or nuclear power? It would seem at last that humans have worked clear of their dependence on fire. Is this so?

Not really. Concerning hydropower, consider the chain of nature that is involved. Rain falls to the earth, it flows into rivers and streams, it collects in lakes and oceans, it is absorbed by the sun, it forms clouds, and the clouds send back rain again. The process has continued since the dawn of time. The answer to the riddle lies in the sun, without which nature's chain of events would grind to a halt. And what is the sun but a huge ball of fire? Not a fire of burning logs, but rather one of thermonuclear reactions which will shine for at least another five billion years.

Nuclear energy plants, monuments to the technical genius of the human mind, recreate on earth the same activity that takes place on the surface of the sun. Just as the philosophers perceived more than two thousand years ago, fire is the force that holds together our solar system.

Unfortunately, man has discovered ways to use fire as a destructive force as well. Let's look at how we have used fire in battle.

4.
FIRE GOES TO WAR

The covered wagons are drawn into a tight circle, and a band of Indians, riding bareback on ponies, gallops 'round and 'round. From time to time an Indian falls to the ground, but this doesn't slow the stream of flying arrows. Then the wagons begin to burn as flaming arrows tear into their canvas tops. The situation is desperate. Haven't you seen the action a dozen times in different movies?

American Indians weren't the first to think of using fire as a weapon of war. Archimedes, a Greek mathematician and inventor who lived 200 years before Christ, is said to have destroyed an invading enemy fleet by using huge mirrors to concentrate rays of the sun on the ships, causing them to break into flames. Nor were the Indians the last to put fire to deadly work.

Perhaps the most terrifying use of fire as a weapon was invented during World War II. The British had

FIRE! ITS MANY FACES AND MOODS

been experimenting with a flammable mixture of gasoline and rubber, but their research was frustrated when Japanese armies gained control of the rubber-producing nations. The United States, however, was successful in developing a jellied gasoline called *napalm*.

Napalm burns with a furious intensity. It can be used in aerial bombs, splashing torrents of flame wherever one lands, or it can be hosed in long fiery streams from flamethrowers carried by infantry and mounted on tanks. Flamethrowers were not new in World War II. They were used by the Germans in 1915 in the World War I battle of Verdun. The horror of napalm, however, is that it clings to flesh, metal, or anything that it touches. Official World War II battle films made no attempt to censor scenes of Japanese soldiers, engulfed in flames, being driven out of fortified island caves.

The first use of napalm bombs came during the war in March of 1945. A fleet of American bombers dropped 1,900 tons of the so-called M69 bombs on Tokyo. The bombs were fused to explode 100 feet above ground level, at which point they scattered thousands of smaller napalm bombs, each burning at a temperature of 3500° C. The firestorm lasted for days, destroying one-fourth of the city, killing eighty thousand people, and leaving a million more homeless.

The use of napalm continued in the Korean and the Vietnam wars. During the years of the latter conflict, peace groups protested what they called the in-

Fire Goes to War

humanity of napalm, and the chemical firm that manufactured it was picketed by protesters on several occasions.

Many other Americans disputed their position, arguing that no weapon's results are pretty and that the goal of warfare is victory. They pointed out that the flame-throwers of today are equivalent to the Roman catapults, which hurled boiling oil into the ranks of the enemy. Finally, fire bombing was not an exclusive tactic of the Allies. We suffered 3,700 casualties at Pearl Harbor, an attack which Japan launched while we were at peace. Hitler's *London Blitz*, which lasted from September, 1940, to May, 1941, was an all-out attempt to batter the English into submission. Much of the city was burned to the ground, and by the time the war ended in 1945, more than thirty thousand Londoners had been killed.

Armies have always used fire against each other, but for a long time the only way of applying it was by use of the torch. In A.D. 146 the African city of Carthage was burned to rubble by the Roman army in the third of the Punic Wars. Fifty thousand inhabitants who failed to escape the city were killed, and when the fire burned itself out the Romans plowed under the remains so that any trace of the city would be removed forever from the face of the earth. Many of the ruins have been uncovered by archaeologists and are a tourist attraction.

During the fifteenth century, Moscow was burned

FIRE! ITS MANY FACES AND MOODS

by an army of Tartars who were attempting primarily to kill just one man, Duke Ivan III. To accomplish this they surrounded the city and ringed it with flames. Not only were the city's buildings wooden, but also the streets were paved with pine logs. The resulting whirlpool of fire drove the inhabitants to the central market area of the city where all 200,000 of them perished.

American cities have not entirely escaped the fires of war. New York City was burned on September 15, 1776, while occupied by the British during the Revolution. The fire was probably started by a band of Colonial extremists, and it was the Tories and the British soldiers who fought to put the fire out. The entire city might have been consumed but for a fortunate shift in the wind that kept the fire on the west side of Broadway. A total of 493 buildings were destroyed, quite a few considering that the city had not yet grown beyond the tip of Manhattan Island.

Washington, D.C., was burned during the War of 1812. Ninety percent of the population had been evacuated when British General Ross entered the city and began setting fire to the public buildings. On the first night he torched the Capitol and the White House. The navy yard also went up in flames, but this fire may have been started by Americans to prevent the British from occupying and using it. The next day, the War and Treasury buildings were set aflame, as well as the arsenal, which stood on Green-

Fire Goes to War

leaf Point. By a twist of fate, a hundred British soldiers were killed and many were injured when flying sparks fell into a well in which the Americans had hidden ammunition. The British pulled out of Washington the following day.

Still a bitter memory for many Southerners is the burning of Atlanta during the Civil War. The city served as a Confederate supply depot, and it was a prime target for the Union Army under the command of General William T. Sherman. The city was captured in September, 1864, and two months later Sherman's troops put it to the torch.

Fire is not always used as a weapon of offense in warfare. It can be used defensively as well. Retreating armies often use "scorched earth" tactics, burning crops, buildings, and anything of value as they run. This deprives the pursuing army of food, shelter, and loot. It requires a great amount of fortitude, however, for a people to burn their own homes and fields.

A dramatic illustration of the effectiveness of scorched earth strategy was provided by the Russian soldiers and civilians who were fleeing from the advancing forces of Napoleon. The French captured Moscow in 1812, but they conquered a city that was abandoned and in flames. Denied sufficient food and adequate shelter, Napoleon's army suffered terribly in the cold Russian winter. After thirty-five days the French, themselves, were forced to retreat westward

FIRE! ITS MANY FACES AND MOODS

and were decimated along the route by surprise attacks from avenging Russian units.

You are certainly familiar with the command "Fire!" given to soldiers in battle. Why not "Shoot!" or some other equivalent command? The answer is simple. In the early years of cannons and muskets, every discharge produced a volcanic eruption of fire and smoke. For the colonists in America, perhaps it was a blessing that the invention of smokeless gunpowder was a long way off. The bell-nosed blunderbuss served as a psychological as well as a physical weapon. Belching flames with each shot, it made hostile Indians think twice before counterattacking, and this provided time for the lengthy process of reloading.

A forerunner of gunpowder known as *Greek fire* was used in the siege of Constantinople in the seventh century. It was a devastating weapon for the era, burning rows of soldiers to a crisp. Used also on ships in naval battles, it was sprayed from bow-mounted tubes shaped to look like snarling beasts. Other boats were equipped with battering rams tipped with barrels of blazing sulphur. The ingredients of Greek fire were probably sulphur or naphtha and pitch, a black gooey tar. Naphtha was also used in India as a weapon. Known as "Oil of Medea," it was murderously effective in battle.

Modern gunpowder was invented in 1242 by Roger

Fire Goes to War

Bacon, a philosopher and scientist. His recipe called for sulphur, charcoal, and saltpeter. The chemical compound potassium nitrate, saltpeter, is a crystallike substance that used to be collected from barnyard-animal droppings in which it forms naturally.

Long before Bacon's discovery, gunpowder was used by the Chinese, the Arabs, and the people of India. Curiously, the Chinese didn't see the full potential of gunpowder to hurl missiles at an enemy. Instead, it was used as a psychological weapon to frighten the foe by shooting off firecrackers, flares, and rockets. To this day the Chinese are masters of pyrotechnic display. On the first day of the Chinese New Year, dragons wind through the streets amid a crescendo of exploding fireworks. The noisy display frightens away evil spirits and assures a happy and prosperous year.

Roman candles that shot flaming pellets into the air were used in the Colosseum for display, but Roman generals never thought of aiming them at an enemy. It wasn't until 1595 that the longbow was replaced by the musket as the standard weapon of the British infantry.

In 1300 a German monk named Berthold Schwarz invented a firearm that used gunpowder to explode shells. The idea spread, and by the middle of the century cannons were in extensive use. The age of impregnable castles and walled cities was coming to an end.

The first major military victory resulting from the

FIRE! ITS MANY FACES AND MOODS

use of the cannon came in 1453 when Constantinople was taken by the Turks. The city was considered to be completely secure, surrounded by thick walls, but sixty-nine cannons battered one section of the wall until it was reduced to rubble. A few years later Charles VII of France smashed sixty castles in sixteen months. In 1523 the strongest castle on the Rhine was flattened in one day.

As soon as castles stopped offering security, the feudal barons lost their power as warriors. Their ironclad horses and suits of armor no longer made them invincible. Now a simple foot soldier with no benefit of armor could bring a knight to the ground with one shot from a musket. As expressed by the historian Carlyle, gunpowder made all men tall.

The projectiles hurled by the earliest cannons were stones; possibly the Arabs used cannons to shoot giant arrows. Only later did the idea develop of a projectile that itself would explode on contact with the target. The trick was to keep such a shell from exploding inside the cannon. In fact, cannons often did explode, killing the gunners rather than the enemy. The Dutch finally invented a way to shoot a bomb out of a cannon without first blowing up the cannon.

Before the introduction of gunpowder, sea battles were nothing more than land battles fought on the decks of ships. When enemy warships sighted each other, the faster one overtook the slower, and seamen lashed handrails together with stout ropes. Then, a boarding party armed with cutlasses leaped over the

Fire Goes to War

rails and fought a pitched battle on the deck and in the shrouds. Meanwhile, archers lined the rail and loosed a stream of arrows at the foe.

Gunpowder changed all that, although a vestige of the past remains on U.S. Navy fighting ships where Marines still form a part of the ship's complement. The first modern sea battle making full use of gunpowder took place when the British navy vanquished the Spanish Armada in Spain's unsuccessful attempt to invade England in 1588. The Spanish wanted to lash ships together in the traditional way and fight with swords. They were infuriated when the British stood off and used cannon volleys in a hit-and-run strategy.

You might wonder how it is possible to shoot an explosive shell without having it blow up inside the gun. The answer is that there are two types of explosives used in modern ammunition. *Low explosives* have a relatively gradual burning action that pushes the shell out of the gun barrel. The shell itself is packed with *high explosives,* which explode instantaneously when the shell strikes its target. The category of high explosives includes dynamite and trintrotoulene, the latter more commonly known as TNT.

If an explosive is not packed in a shell or some other casing, it will merely burn when ignited. If you break open a firecracker and put a match to it, it will flash but not explode.

When an artillery shell strikes its target, the im-

FIRE! ITS MANY FACES AND MOODS

pact sets off a fuse inside the shell. The fuse can detonate the explosive charge immediately, or the reaction can be delayed if the shell is intended to pierce armor before exploding. Time- or radio-activated fuses can cause a bomb to explode before it strikes the ground, just as antiaircraft shells can be made to explode in mid-air and scatter shrapnel. Bombs can be fused so that the explosion is delayed for hours or days after they have struck the ground. During the London Blitz in World War II, posted signs warning of unexploded bombs were a common sight. It required a great deal of courage for bomb crews to deactivate these bombs because the smallest slip could mean death. Also, some of the bombs were booby-trapped to go off the moment someone tried to open them.

Infantry rifles and handguns project lead slugs rather than explosive missiles. Their firing principle is the same as that of the Pilgrim muskets, but in those days powder, wadding, and shot had to be rammed down the barrel one by one. Firing was achieved by applying a spark from a flint, which ignited powder in the firing plate. As early as 1700, cartridges made of charge and ball were wrapped individually in paper. Brass and steel casings were introduced in the late 1800s, after the Civil War, along with breech-loading rifles and smokeless powder.

Nuclear power has now made all conventional weapons as antiquated as broadswords and battering

Fire Goes to War

rams. Human genius has evolved to the point where we can now destroy the planet on which we live. A far cry from flaming arrows!

Destruction by fire through carelessness and accident has occurred frequently over the years. The next chapter describes some of these famous disasters.

5.
FIRE DISASTERS

Can you name the three fires in the history of our country that were the most tragic in terms of lives lost?

Your first guess might be the Great Chicago Fire of 1871. The earthquake-fire of San Francisco in 1906 would be another likely selection. Many books and movies have told the stories of both these events. In fact, the Chicago Fire ranks eleventh, with a loss of 250 lives. The San Francisco holocaust ranks fifth, with its toll of five hundred victims.

According to the Insurance Information Institute, the three most deadly fires were those described below—events that you have probably never even heard of. Here is a brief account of each.

A Night of Judgment

The night of October 8, 1871, was a bad one for the city of Chicago. Fire broke out in the barn of the O'Leary family, who lived at 137 De Koven Street.

Fire Disasters

Whether their cow started the whole thing by kicking over a lantern, as legend has it, is of little concern; the resulting fire engulfed almost the entire city of wooden structures, claimed 250 lives, left 100,000 persons homeless, and destroyed 17,500 buildings.

Newspapers throughout the world told the story. Not a single paper, however, made mention the following day of another fire that broke out the same night only 250 miles north of Chicago. This second fire was even more furious than the Chicago Fire: it claimed more lives, and it destroyed a far greater area. The inferno received little outside attention, however, because few people knew about the small logging town of Peshtigo, nestled in the vast Peshtigo Forest of Wisconsin.

The summer and early fall had been hot and dry, always a dangerous combination in timberland. Trees and ground covering were as brittle as kindling, streams were low, and wells were almost dry. From the pulpits of the town's two small churches the preachers, almost by a gift of revelation, warned of a coming judgment day.

On a Sunday evening the forest burst into flames. Perhaps it was started by a bolt of heat lightning. Maybe a spark was produced by the friction of two trees rubbing together in the wind. There is no way of knowing for sure. Whatever the original cause, the town of Peshtigo as well as several neighboring communities were rapidly trapped in the center of a hurricane of flames.

FIRE! ITS MANY FACES AND MOODS

The fire was so violent and the heat so intense that there was no chance of fighting back with water. The only hope lay in escape, but this became impossible when rising winds engulfed the entire region in gales of superheated air. Buildings exploded into flame without any visible source of ignition. Even people burst into flame as they ran wildly through the streets. One man sought escape by jumping into a trough used for watering horses. He was boiled to death.

On a bridge that crossed the river, masses of people tried to escape in both directions. In a state of panic they fought each other until the bridge collapsed, drowning many of them. Even many of those who sought safety in ducking beneath the surface of the river were lost because they were incinerated in the split second it took to raise their heads for a gulp of air.

On the following morning nothing remained but black ashes, which included the bodies of 1200 victims. Before the fire finally spent itself, 4 million acres of prime forest were destroyed, an area roughly the size of Connecticut and Rhode Island combined.

By some miracle, a few people survived. They rebuilt Peshtigo, and on the land that had been cleared by the fire they planted crops. Today when you look upon the peaceful dairy farms that cover the region, it is difficult to imagine what it must have been like on that night of judgment.

Fire Disasters

Theater of Death

When you go to the movies you probably never think for a moment about fire or of being burned and trampled to death. Safety regulations govern theater construction and management, and you are probably safer in a theater than you are in your own home. Your peace of mind, however, has a high price. You are in debt to 602 persons, many of them children, who lost their lives less than a week after Christmas in the year 1903.

The victims were among an audience of two thousand who jammed the Iroquois Theater on a frigid Wednesday afternoon in Chicago. The brand-new showplace had opened only a month earlier, and it was the most beautiful in the Midwest. Moreover, it was fireproof. At least that is what the city's building commissioner said.

Not everyone agreed with him. A few days before the tragedy a stagehand had complained that the asbestos curtain separating the stage area from the audience did not lower completely because it was blocked by a lighting reflector. His complaint was shrugged off because there were more important things to think about.

Also ignored were several criticisms made by the editor of *Fireproof* magazine. There was no sprinkler

FIRE! ITS MANY FACES AND MOODS

system over the stage. The skylight above the stage was nailed shut, allowing for no ceiling vent to carry flames and smoke upward should fire break out. There was no fire alarm system linked directly to firehouses.

If only someone had listened.

The theater's 1,724 seats were filled, and three hundred additional patrons paid for standing room. The second act of *Mr. Bluebeard* was just starting, and the stage was alive with pretty dancing girls in fetching costumes. Eddie Foy, a famous comedian and the star of the show, was in his backstage dressing room getting ready for his entrance on stage. High above the stage, out of the audience's view, a stagehand aimed a carbon-arc spotlight on the dancing girls below.

Old-fashioned carbon lamps generated tremendous heat. When a flimsy bit of scenery brushed against the lamp, the fabric ignited.

Every fire starts small. For a moment the flame was no larger than that of a match. The stagehand tried to clap it out with his hands. But this failed, and the lick of flame began to climb. Now the stagehand reached for a fire extinguisher, but its spray fell short by just a few inches. Then it was too late.

Flames leaped along the paper scenery, and flaming debris began to rain down on the performers below. Eddie Foy heard the commotion it caused, and he rushed to the footlights to plead with the audience to remain calm. His words might have had good effect, except for three things.

First, the asbestos curtain failed to drop completely to the stage floor. Second, someone backstage opened an exit for the showgirls to escape. This caused a backdraft, and with no vent above the stage, a huge blowtorch of flame bellowed out into the audience. Third, falling scenery knocked over the electric switchboard, and the theater was plunged into darkness.

The panic was beyond imagination. Women and children were trampled. People fell from the balcony into a pit of flames. Those who reached the upper fire escapes found that there were no ladders to the ground. Lower-level exits opened onto an unexpected four-foot drop, and people fell, breaking bones as they piled up on each other. Some of the fire-exit doors opened inward, and those who reached the doors couldn't open them because of the pushing from behind. In the main lobby the bodies became so intertwined that later they could not be separated.

Firefighters took only fifteen minutes to control the blaze, but by then the theater was a blackened shell littered with charred corpses.

An inquiry uncovered a long list of violations, and several individuals were indicted for negligence and manslaughter. But no one ever went to jail, and not a single penny was paid in damages to the injured or to the families of those who died.

FIRE! ITS MANY FACES AND MOODS

But even tragedy has a way of serving us. As a result of the Iroquois Theater Fire, cities everywhere embarked on a crusade to enforce safety regulations. Theaters today have automatic sprinklers and quick-opening vents above their stages. Curtains and settings are truly fireproof. Back-up emergency lighting systems are required. Exit doors are clearly marked, and they must open outward. The list goes on and on.

Along with other theatergoers, you enjoy the safety that was paid for more than seventy-five years ago.

The Fatal Outing

The horror of fire is always intensified when it claims women and children as its victims. The fateful family outing sponsored by St. Mark's German Lutheran Church in New York City turned into the second most dreadful fire in the history of our nation. Before a single picnic basket was opened, 1030 persons were dead, almost twice the toll of the Iroquois conflagration. Of those who were lost, 750 were babies and young children.

June 15, 1904, was a beautiful sunny morning, and throughout the German community husbands at breakfast tables teased their wives and children about having a day of fun while they had to go to work. More than one hundred of these husbands would never see their families again.

The outing started at the church with a happy

Fire Disasters

parade through the streets, the melodies played by a German band mingling with the laughter of the children. The parade led to the East River pier at Third Street, and then it took almost an hour for the 1,358 picnickers to file aboard the excursion boat *General Slocum*.

The ship was a big three-decked sidewheeler that could have held twice the number of passengers. It had been freshly painted, and brightly colored banners flapped briskly in the stiff breeze. The plan was to head north on the East River and then out to Locust Grove on Long Island Sound.

The captain was William Van Schaick, a mariner with forty years' experience, thirty of them as a licensed captain. Despite his credentials, however, Van Schaick was to become the villain of the day.

A boy knocked on the door of the pilot house about the time that the boat had reached Ninety-second Street. He wanted to report that he had seen smoke coming from a stairwell that led down from the main deck. Reports say that Van Schaick ignored the warning, telling the youth to get lost. Meanwhile, the fire, which started in oily rags stored too close to a kitchen stove, began to spread. A woman on deck walked over an open grate, and her clothing burst into flames from the heat of the fire below. She screamed, and panic took over.

The captain ordered his crew to man the hoses, but there were two problems. The crew had never

FIRE! ITS MANY FACES AND MOODS

run through a fire drill, and even if they had, it would have served no purpose because all the hoses were split from lack of maintenance.

If Van Schaick had turned to shore and signaled for help, the day might have been saved. But for some reason the captain didn't do this, continuing instead upriver at full speed. This created a wind that fanned the flames to a blast furnace intensity. Still, he was determined to ground the ship on mudflats, which were three miles distant.

Flames and smoke were everywhere, and mothers in desperate search of their children ended up fighting each other. Many jumped overboard and were mangled when they were sucked into the huge sidewheel. Others drowned in the churning wake of the fast-moving ship. By now the *General Slocum* was being trailed by a fleet of small boats wanting to help but unable to do so. Only one person in ten who jumped could be saved.

The ship, a flaming torch, finally ground into the beach of North Brother Island. The impact caused the flame-consumed upper decks to collapse, trapping anyone still alive below. Those who could escape walked to shore over a carpet of bodies floating in the shallow water.

In the inquiry that followed, an incredible litany of negligence and indifference was recited. The ship had passed an inspection the day before the tragedy, and the inspector had failed to notice that the lifeboats were locked by rust into the davits, impossible

Fire Disasters

to launch. Life preservers were wired to the boat, and those which could be torn loose crumbled into dust. The hose lines were never checked, and none of them were operable.

Eleven persons, including ship company officials, were brought to trial. The only one sent to jail was Captain Van Schaick. He was given one to ten years in Sing Sing, and was pardoned and paroled six years later by President William Taft.

Once again, lessons were learned too late. Today, every waterfront city has a strict maritime code, and the Coast Guard rigidly enforces all Federal regulations. Another *General Slocum* would be impossible now.

These three fires took place quite a long time ago, but similar tragedies can happen today, even though the probability is reduced by fire codes and modern firefighting methods and equipment. The sixteenth worst fire in our history took place just a short time ago.

On May 28, 1977, fire destroyed the Beverly Hills Supper Club in Southgate, Kentucky, leaving 165 persons dead and more than one hundred injured. The disaster would have been greater but for an eighteen-year-old busboy who saved hundreds of lives by leading hysterical patrons and employees to fire exits.

For many, the fire brought back the memory of the Coconut Grove fire, another famous nightclub conflagration which took place in Boston in 1942. The

FIRE! ITS MANY FACES AND MOODS

country was at war, and many of the 492 people who perished were servicemen having a last party before being shipped overseas.

Modern times give rise to modern hazards. In the summer of 1978, six hundred tourists had parked their campers and pitched tents on the edge of a lake near the Spanish resort town of San Carlos de la Rápata. It was midafternoon and children romped on the beach and splashed in the water as their parents prepared lunch on butane stoves.

On a long bend of highway above the camp, an occasional car passed by unnoticed. A huge 38-ton tanker-truck might have passed unnoticed too. It was going only 40 miles per hour, but for some unknown reason it skidded out of control, crashed through a retaining wall, and rolled over. Even at that point the accident could still have been minor. But the tank was carrying propylene gas. It exploded and engulfed the campsite in a sea of flames. Automobile gas tanks and cylinders of butane set off a series of secondary explosions. In a very short time, one hundred persons were dead, and within a few days the death toll reached 150.

Tank cars can't be banned from highways and railroads; we are too dependent on the benefits they provide. It makes you realize, however, that we should all think more about safety rules and regulations. How many of us live close to highways and railroads that represent similar hazards?

Fire Disasters

We can't prevent all fires, but we can learn from them and prevent history from repeating itself. Hospitals today don't use nitrocellulose film for X-rays because of a 1929 fire in a Cleveland hospital where 121 persons died from poisonous smoke. It is now against the law to lock school doors while school is in session as the result of a 1908 fire in Collingwood, Ohio, where 175 children under the age of fourteen were killed. Fire escapes and safety exits are now required in factory buildings because 175 women and girls died in New York's Triangle shirtwaist factory fire in 1911. The same fire led to the creation of the New York Department of Fire Prevention.

In July, 1944, a huge circus tent caught fire in Hartford, Connecticut, and 164 members of the audience perished. The tent had been treated with a mixture of bees' wax and kerosene to make it waterproof! And because in 1947 the law didn't require cargo labeling, 568 were killed in a catastrophe in Texas City, Texas. Fire fighters had no way of knowing that the hold of the burning S.S. *High Flyer* was loaded with highly explosive sulphur and ammonium nitrate.

You'll find many books on your library's shelves telling about other famous fires in history. This chapter can't include them all. But one recent fire deserves to be mentioned because it ranks among the saddest, even though only three people died. Can you guess which fire it was?

Hardly a person alive failed to feel a sense of loss over the deaths of three of our astronauts who gave their lives inside their Apollo spacecraft capsule. They were Lieutenant Colonel Virgil Grissom, 40, Lieutenant Colonel Edward White, 36, and Lieutenant Commander Roger Chaffe, 31. It is ironical that they didn't die in outer space where the hazards are so great, but in their craft while it was docked securely to the ground.

After ten flawless Gemini trips into space by several different astronauts, the three were preparing for the next launching. They were strapped into their reclining launching seats inside the capsule for a simulated launch countdown. The cabin was pumped full of oxygen; just then a short circuit produced an electric spark, and the interior burst into flame and smoke. The three astronauts' suffering lasted less than fifteen seconds. They died in the service of their country. The date was January, 27, 1967.

Long ago, in A.D. 79, another fire-related disaster buried a whole city. The city was Pompeii, and the fiery fury was caused by a volcano.

6.
VOLCANO

Volcanoes were once considered rare phenomena in the United States. Since the 1959 admission of Hawaii as our Fiftieth state, however, this is no longer true. The island's wonderland of volcanoes became Hawaii National Park in 1916 by an act of Congress, and in 1961 the Haleakala section became Haleakala National Park. Two of its peaks, Kilauea and Mauna Loa, are among the world's most active volcanoes. Whenever an eruption is anticipated, thousands of people crowd the scene to witness one of nature's most incredible displays of fire's might.

Native Hawaiians say that the fire goddess Pele lives in the depths of Kilauea, and they drop offerings into the smoking crater to stave off her anger. A long time ago these offerings even included an occasional human sacrifice.

Kilauea's most recent eruption took place in September, 1977. For several days the mountain spouted fire and tossed lava bombs, rocks, and ashes 350 feet

FIRE! ITS MANY FACES AND MOODS

into the sky. A river of red-hot lava 40 feet deep and 1,000 feet wide flowed down the slope at the rate of 1,000 feet per hour. The potentially lethal flow of melted rock stopped just 400 yards short of consuming the small village of Kalapana.

What causes such a spectacular natural event? No one is absolutely certain, but scientists are confident that they have a sound theory. To understand it you have to know about a geological discovery called *plate tectonics.*

Despite appearances to the contrary, the surface of the earth is not one solid and continuous mass. Instead, the crust of the earth is broken into several vast plates, separated from each other by enormous fissures or cracks. The plates drift eternally on a sea of melted rock, moving a few inches a year.

Look at a map of the world and notice how the continents of Africa and South America could fit together like jigsaw pieces. Based on the plate tectonics theory, the two continents were once united, but over many millions of years they have drifted apart. Amazingly, an underwater range of volcanic mountains bisects the Atlantic Ocean in a north-south direction, and the configuration of this range is identical to that of the separated continental coastlines. This suggests the possibility that the two continents were once separated by a large crack, which spread wider and wider as a result of volcanic activity. Further support

Volcano

of this theory comes from the fact that all ocean bedrock is volcanic in origin.

The tremendous pressure resulting from the rubbing together of the earth's plates causes heat, and this results in pools of *magma* or melted rock beneath the surface. The magma, combined with large amounts of gas, rises and forms vast reservoirs very close to the earth's surface. Upward pressure sometimes forces a channel to the surface, and when this happens the result is a volcanic eruption.

Volcanoes almost always occur along plate boundaries, except in midocean where the crust of the ocean floor is very thin. Exceptions to this rule, which puzzle geologists, are the several midcontinental volcanoes in Africa.

The first stage of a volcano is the eruption of flaming bellows of gas. This is followed by the ejection of dust, ash, and large fragments of stone called volcanic bombs. Lava, with a temperature of 1100° C, often flows down the slopes, but lava flow does not always occur. The city of Pompeii was buried entirely in dust and ash. The 1883 eruption of the island of Krakatoa in Indonesia blew dust 17 miles high, and the cloud of dust circled the earth several times, causing brilliant red sunsets in many parts of the world.

Volcanoes don't really come from a mountain. It is more correct to say that they form a mountain. Each eruption deposits a new layer of volcanic debris, and

FIRE! ITS MANY FACES AND MOODS

these layers build up over countless years to form a mountain. *Compound volcanoes* spew out both ash and lava, and the combination results in a hard-crusted mountain. If there is no lava, a *cinder cone* is formed. In 1943 a crack appeared in a cornfield in western Mexico, and cinders shot into the sky. By the time the eruptions ended in 1952, a cinder cone had formed to the height of almost 1,500 feet.

Most of the world's volcanic activity is unseen, taking place on ocean floors. Over vast stretches of time these volcanoes have formed underwater mountain ranges whose peaks would dwarf the mountains on dry land. Sometimes peaks extend to the water's surface and above, and thus volcanic islands are formed. At the northern tip of the subatlantic range, Iceland is such an island.

You might wonder how Icelanders feel, knowing that they are living on a volcano. Many of them consider it an advantage, particularly in this age of energy crisis. In the city of Reykjavik the residents don't worry about supplies of heating fuel. They heat their homes with water piped from the hot volcanic springs that flow a few feet beneath their feet. Still, from time to time they must pay for the convenience, the most recent installment being the eruption of 1973.

It began on a January morning at two A.M. on Heimaey, the largest of the islands off Iceland's western coast. The earth cracked open dangerously close to a village of five thousand people, and fountains of lava and flame jetted hundreds of feet into the air.

Volcano

Volcanic bombs skyrocketed and burst in midair, and glowing lava flowed downhill toward the harbor. Sections of the island were buried in ashes. Tragedy was averted only by the evacuation of most of the inhabitants to the mainland.

Today, tourists to the island find flowers growing out of the ash, and steam-spouting chimneys still extend from heaps of solidified lava. In some places you can find enough heat to cook a meal just by scratching the surface of the ground.

One of the largest of the earth's plates, the Pacific Plate, which circles the Pacific Ocean and borders on part of the west coast of the United States, accounts for the mountain formation of the western states. Lassen Peak in California erupted in 1921, but no eruptions have taken place on the mainland of the United States since then. An Alaskan eruption took place on Mount Katmai in 1912, before the territory became a state. There is virtually no possibility that a volcano could erupt anyplace in the United States other than in the regions already mentioned.

Volcanoes are classified as *active, intermittent, dormant,* or *extinct.* Active volcanoes, such as Stromboli on the island of the same name off the coast of Italy, are in a state of constant eruption. Hawaii's Kilauea is called intermittent because it erupts at regular intervals. Lassen Peak is a dormant volcano in California. Dormant means "asleep," but there is no guarantee that such a volcano will not reawaken some day. Extinct volcanoes are those that have been

FIRE! ITS MANY FACES AND MOODS

inactive since the beginning of recorded history. Kilimanjaro in Africa, familiar as the setting of a short story by Ernest Hemingway, is an extinct volcano.

People who live in the neighborhood of volcanoes are fairly safe today because such mountains are constantly monitored, making it possible to forecast eruptions. Not too much can be done to protect property, but lives can be saved by timely evacuation. The people of Martinique, an island of the West Indies, were unprepared, however, when Mont Pelée erupted back in 1902. The pressure was so great that the mountain blew apart, and before the eruption ended 29,000 persons had perished.

In 1815 the Indonesian volcano Tambora took 66,000 lives. Ten thousand were buried in the ashes, and the remainder died from famine and epidemics triggered by the eruption. The greatest tragedy of all was the 1883 blowup of Krakatoa, also in Indonesia. The force of the eruption reportedly could be heard 3,000 miles away, roughly the distance from Los Angeles to New York City. Sea waves more than 100 feet high were created which swamped many neighboring islands, drowning 96,400 inhabitants. In the past 250 years, close to 250,000 people have been victims of vocanic wrath.

History's most famous volcanic eruption was that of Italy's Mount Vesuvius in A.D. 79, which buried the town of Pompeii beneath a thick blanket of volcanic ash. On the beginning of the second day of activity,

Volcano

the sunrise never came because the sky was so black with dust.

In the early years following the catastrophe, shafts were dug down into the ruins by looters in quest of valuables. It wasn't until the late 1880s, however, that scientific excavations began to uncover the splendor and decipher the way of life of an ancient civilization.

The excavations are still in progress, and visitors today are able to stroll along streets flanked with the remains of shops, houses, and villas. The graffiti on the walls are still legible.

The ruins reveal that Pompeii was a city of about twenty thousand. Most managed to escape, but about one of every ten died, many because they couldn't bear the thought of leaving their belongings behind.

It is obvious that the eruption was unexpected. Evidence indicates that normal life was enjoyed up to the final moment. Archeological digging has uncovered houses in which food was laid out on the tables —meals that were never eaten.

Bodies of many victims were buried in ash, and after the ash hardened and the bodies decayed, perfect molds were formed. By filling these cavities with plaster, many hundreds of casts of human and animal fossils have been made, providing a grim recreation of their death agonies. The contorted positions of the victims indicate that many died from the poisonous gases that spewed from the mouth of the volcano. Also, it appears that human nature wasn't too different than from what it is now. Some people were still clutching bags of gold coins they were carrying in their unsuccessful attempt to get away.

The fate of a neighboring city, Herculaneum, was different from Pompeii's. It was buried not in loose ash, but in a flow of mud that formed when heavy rain mixed with volcanic dust. Archaeologists have found many priceless manuscripts and works of art in the ruins, but the digging is much more difficult than at Pompeii because of greater depth and the solidity of the hardened mantle of mud. A third city, Stabiae, was also claimed by Vesuvius' fury.

Interestingly, the heavy rain that fell immediately following the eruption was probably induced artificially by atmospheric reaction to the burning gases. A similar effect occurred in modern days when the German airship *Hindenberg* exploded and burned while making a landing at Lakehurst, New Jersey, in 1937. The rescue operation was made easier by the sudden rain that wetted down the burning wreckage.

Books and movies have told the story of Pompeii many times, but many Americans became engrossed in this event only recently when an exhibit of the treasures of Pompeii was displayed in Boston and New York museums. Thousands stood in line to get a look at the art and artifacts, the rooms and their furnishings, which were left behind the day of the disaster.

You should not conclude from this chapter that volcanoes are all bad. Quite the contrary is true. Volca-

Volcano

noes are a vital part of nature's scheme, and the world would not be such a good place without them.

Think for a moment about the huge plates that unite to form the surface of the earth. Without the force of volcanoes to help shift these plates, all the land masses of the world might still be concentrated in one gigantic continent, just as they were billions of years ago. The inner region of this supercontinent, being so far removed from the sea, would be a desert waste. Only coastal cities would have the economic benefit of shipping routes. Indeed, civilization would be confined to a very narrow strip surrounding the continent.

Other, more immediate benefits are derived from volcanic activity.

Soil

The initial result of a volcanic eruption is the destruction of all vegetation for miles around. This results from the suffocating blanket of dust and ash that falls to the ground. These volcanic materials, however, are ultimately broken down by sun, wind, and rain, and they form a soil that is rich in potassium, which is essential for plant growth.

The greatest population concentrations in Indonesia, a country with generally poor soil, are in the rich volcanic regions. Coffee from Costa Rica and Guatemala, said to be among the best in the world, comes from plants growing on volcanic slopes.

FIRE! ITS MANY FACES AND MOODS

Even lava flows that become brick-hard ultimately break down into excellent tillable soil.

Commercial Products

The most common commercial product created by volcanoes is *pumice*. You can buy lumps of it in most pharmacies, as its abrasive surface can be used to rub away callouses and to massage the skin. Sometimes it is ground into powder to be used as a gentle abrasive in toothpaste, although limestone is the more commonly used agent.

Pumice is formed when liquid lava cools very rapidly under little pressure. Dissolved gases escape, leaving countless tiny holes and giving the stone a spongelike texture.

Farmers on the Canary Islands in the Atlantic have a plentiful supply of pumice, so they spread it on their fields. It keeps out the direct heat of the sun, yet allows water to filter through, and weeds don't get a chance to take root. Many of the native islanders carve their homes directly into the base of pumice cliffs.

In Peru pumice is cut into building blocks called *sillars*. The blocks are light and easy to handle, and they can be cut and shaped with a hand saw. Yet they are strong, and houses made from them are comfortably dry because of their ability to absorb water. In Naples pumice is ground into *pozzolana*, an excellent cement that has been used since the days of Julius Caesar.

Volcano

Before the age of steel, the world treasured the keen knives that can be shaped from a form of black glass found in lava flows. The Lipari Islands off Sicily were once a center of trade because of their natural supply of this material. When Spanish explorers discovered the Aztec, Mayan, and Inca civilizations of South and Central America, they found that knives made from this material were commonly used in the ceremonial human sacrifices.

Ores

Gold, silver, copper, sulphur, and other minerals would be almost nonexistent on the surface of the earth without the action of volcanoes. The rich mineral deposits of the world would still be buried deep in the center of the earth.

The most productive mining operations of North and South America are located along the volcanic edge of the American plate. Cerro Rico, an extinct volcano in Bolivia, is one of the world's greatest sources of silver. The soil of the interior of the United States contains only three parts of copper per million. In certain volcanic regions of the southwestern United States, however, the proportion is ten thousand times as great.

Power

In a speech to the nation in the summer of 1979, President Jimmy Carter declared an energy war. He pledged to initiate a program aimed at making our country independent of foreign fuel sources. The president spoke of the potential of our own energy resources, but he didn't mention what could be an exciting possibility for the future—geothermal power.

No doubt you know about Old Faithful, the spectacular geyser that performs hourly in Yellowstone National Park. Did it ever occur to you that Old Faithful's power could be put to work? It will never happen to Old Faithful, of course, because the geyser is more profitable as a tourist attraction than it would be with pipes stuck into it to draw off heat. But the idea of geothermal power, dating back to 1904, holds many possibilities. Electricity is now being produced commercially at The Geysers in California, and operations are being expanded. There are many regions where a temperature of 350° C is reached a little more than a mile below ground level. By piping water to a sufficient depth, steam could be produced to run turbines and produce electricity. A giant geothermic electric-generating plant operates in New Zealand, and Russia has been experimenting with a similar plant.

It is estimated that the energy spent in a single eruption of Hawaii's Mount Kilauea would supply the entire nation with electricity during the period of eruption. In addition, such energy is *clean* energy, in contrast to nuclear energy, which produces dangerous waste.

The technological challenge is great, and the cost would be high. But if our country is capable of putting

Volcano

men on the moon, isn't it capable of victory on other scientific fronts?

For future generations the terror of volcanic destruction might well become the blessing of volcanic power.

A large forest fire can wipe out thousands of acres of beautiful woodland and wildlife in just a few hours. The next chapter describes measures for fighting and preventing forest fires.

7.
FOREST FIRE

At the end of a long, hot, dry summer, your thoughts probably turn to the oncoming football season and winter sports. The thoughts of forest rangers, however, are nowhere near so comfortable and relaxed. For them it is the most dreaded season of the year—the time for forest fires.

There are 630 million acres of standing timber in the continental United States, not including the vast woodlands of Alaska. Every year five million of these acres are lost in an average of three hundred fires a day.

In 1891, President Harrison set aside the Yellowstone Timber Reserve, comprising more than a million acres of timberland in northwest Wyoming; since then, the Federal Government has acquired about one-fourth of the country's timberland to insure the well-being of future generations. Forest rangers, under the supervision of the Department of Agriculture, have the responsibility of policing these vast stretches,

Forest Fire

just as fire wardens are responsible for timberland that is owned by individual states. Their greatest enemy is fire.

Forest fire means more than the loss of wood that could have been used to build homes and furniture or to make pulp for newspapers, books, and magazines. Also lost is the wildlife that populates the forests, and the watershed that feeds streams and rivers with rain water. And how can you put a price tag on the lost natural beauty and recreational potential? A forest which is destroyed in a matter of hours might take a hundred years to regrow. Ninety percent of all forest fires are the result of human carelessness.

The most important strategy in fighting forest fires is to spot them before they get a chance to spread. A thimbleful of water can save a million trees if it can be gotten to the spark that is the inception of every raging inferno. Optimistic perhaps, but rangers try to come as close to this ideal as possible.

The first line of defense is a system of lookout towers placed at regular intervals throughout our national forest system. If you suffer from acrophobia, the fear of being in high places, you would not want to be on top of one. Built on hilltops, the towers are 60 feet tall, and from these perches rangers can look down on the forest for miles in every direction. Atop each tower is a glass-enclosed room, 14 feet square, surrounded by a catwalk.

Imagine for a moment that you are on fire lookout

FIRE! ITS MANY FACES AND MOODS

and you spot a small wisp of smoke rising from some distant point. What do you do?

First you convince yourself that it isn't a permanent source of smoke, such as a sawmill or a logging camp. You go into your airborne cabin and step up to the *fire finder*, which is mounted in the center of the room. This consists of a telescope with hairline sighting, set on a disc that indicates the 360° circular division of the compass. True north is at 0°, with other headings measured clockwise from north. East is 90°, south is 180°, and west is 270°. By sighting on the smoke with the telescope, the exact compass direction of the fire is determined. Then you telephone the information to headquarters.

One problem remains. You know the direction of the fire, but you have no way of measuring exactly how far away it is. Fortunately, you don't have to guess. On other hilltops are other towers, and spotters on these are doing the same thing that you have done. When all the spotters in the area have reported their individual sightings, lines are drawn on a map at headquarters, and the intersection of the lines gives the exact location of the fire.

This information is relayed to the nearest shock-troop camp. During dry seasons "hot shot" crews are on twenty-four-hour duty, and within minutes one of these crews is on the move. Usually they travel by truck, but if the region is remote they might be transported by helicopter. They carry only basic tools like axes, shovels, and backpumps. Many fires never make

Forest Fire

the headlines because these shock troopers get to them in time.

If the fire already has a start, the crew radios for help and concentrates on containing the flames until reinforcements arrive.

Help doesn't come in the form of wailing engines. City pumpers can't wind through woodland trails, and there are no hydrants to hook into anyway. More practical are jeeps and bulldozers, the latter called "cats" because of their Caterpillar tracks.

Forest fires are fought with muscles and hand tools, and the strategy is different from that used in fighting city fires. There is not enough water to put out the fire by cooling it. Instead, it must be extinguished by depriving it of fuel. Depending on the size of the fire and the density of the woods, a band of sufficient width has to be cleared in the path of the flames. It is hoped that when the fire reaches this cleared belt it will go out for lack of additional fuel. This is called setting up a fire line. On the unburned side of the fire line, patrols are deployed to put out spot fires resulting from flying embers.

If it is a raging fire, the fire line has to be very wide in order to stop it, and this is when the cats go to work, ripping up brush and tearing down trees. Everyone works under the command of the fire boss who decides where the fire line should be laid.

When even the widest fire line of cleared land seems too narrow compared to the size of the fire, the tactic of *backfiring* might be called for. This involves setting

FIRE! ITS MANY FACES AND MOODS

a new fire along the edge of the control line. When it is done expertly, the new fire burns toward the oncoming fire, and when they meet both burn themselves out.

Backfiring is always a gamble. An unlucky change in wind direction can result in spreading the original fire, and there is always the danger of trapping fire fighters between the backfire and the original blaze. In order to start a backfire, rangers use torches and backpack flamethrowers.

Just as military operations have become modernized, so too have the techniques of fighting fires in the timberlands. Air power is the most significant fire-fighting development since the 1940s. Scout planes spot fires early, and helicopters can land men in places inaccessible on foot. Once a fire has made headway, bombers swoop over the path of flames dropping a creamy mixture of borate and water, with red dye added to make it visible. This solution is a flame retardant, and it clings to the trees in the path of the fire, remaining wet for hours.

If a lake is nearby, helicopters dangle large drums from steel cables and ferry back and forth to the fire, dipping up 900-gallon scoops of water and dropping them on the flames. Pilots have to be highly trained, because low-altitude flying and contending with thermal updrafts calls for great skill.

Often paratroopers play a role in the battle. "Smoke jumpers" are outfitted with coveralls, safety helmets,

Forest Fire

and steel-mesh face masks. Anywhere from two to seventy-five men might be dropped to handle a troublespot. On the ground their efforts are coordinated by radio or walkie-talkies.

You might wonder what you would do if you were a smoke jumper and your chute got snagged on a limb high above the ground. In your pocket you would have a strong, thin line, and, like a mountain climber, you would rappel to the ground by attaching the line to a branch and then passing it under one thigh, across the body, over the opposite shoulder, and then sliding down.

The most promising weapon of the future for fighting forest fires is the computer. By feeding into a computer pertinent information such as wind speed and direction, ground temperature, the type of timber burning, and other data, the computer will be able to predict the fire's behavior and indicate the best strategy for combating it. In the end, however, the conflict will still have to be won by brave woodsmen slinging shovelfuls of dirt and never knowing what hazard the next minute might bring. For example, volunteers fighting a midsummer fire in 1979 in a remote area of Idaho were suddenly faced with countless rattlesnakes driven from cover by the flames.

American Indians are among the most respected of all forest-fire fighters. The first Indian crew was formed in 1949, and since then many such crews have been organized among men from the Navaho, Hopi,

FIRE! ITS MANY FACES AND MOODS

Apache, and other nations. Most of them are small and quiet, but by blood and instinct they are giants in the forest, and when they take positions along a fire line the feeling gets around that everything is under control.

Like hurricanes, when forest fires get big enough they are given names to be remembered by, and campfire talk often revives memories of the many battles that have been won. If you would like to get an idea of what it's like to be in the middle of forest-fire warfare, you'll find in your library a great novel that was written about thirty years ago. The name of the book is *Fire*, by George R. Stewart. Set in the Sierra Nevadas, the book tells the story of a fire named *Spitcat*; it is written with authority, capturing the grandeur and fascination of one of nature's oldest struggles.

Whereas forest fires are caused largely by accidents or carelessness, *arson* is the intentional setting of fires. Read on and discover why it is considered truly a ruthless and costly crime.

8.
CRIME WITHOUT PITY

On April 10, 1979, the U.S. Navy carrier *John F. Kennedy* lay docked at Portsmouth, Virginia. On that day eleven shipboard fires broke out in a five-hour period. One sailor was killed, and thirty-four were injured. The fires weren't caused by poor construction or faulty maintenance; they had been set deliberately.

When you think of crime, you probably think of robbery, kidnapping, or murder. If you were asked to list the crimes that threaten society most, chances are you wouldn't think of including arson. Nevertheless, the intentional setting of fires poses a grave threat. The incidence of arson has more than tripled in the past ten years, and it is the fastest-growing crime in America.

In a typical recent year more than 100,000 cases of arson were recorded, and they resulted in one thousand deaths and countless injuries. On an average night in New York City ten fires are started on purpose, and during the power blackout in July, 1977, one thousand

FIRE! ITS MANY FACES AND MOODS

fires were set in a two-day period. Property damage resulting from arson exceeds $3 billion per year. It costs another $12 billion in indirect economic losses such as unemployment, lowered production, medical costs, and much more.

The state of Missouri has disclosed that half of all fire losses in the state result from arson. The city of San Diego has counted 557 incidents of suspected arson in a ten-month period. In any city or state in the nation the story is the same. If the trend continues, the national loss will soon reach the level of $5 billion per year.

Who pays the bill?

Don't make the common mistake of thinking that only the insurance companies take a beating. It is true that they write out the checks, but their money comes from the premiums paid directly or indirectly by you and every other American. Out of every premium dollar paid for fire insurance, forty cents pay for damage that results from arson. Because of this every homeowner pays inflated premium rates. If you rent your home, the rent is higher than it would otherwise be because it includes your share of the high rates paid by your landlord.

Cities are beginning to take action to fight arson. Seattle was one of the first cities to organize an arson task force. Started in 1974, this force uses the combined efforts of municipal government, prosecuting attorneys, police and fire departments, and the public. A telephone hotline has been established, and the

Crime Without Pity

public is encouraged to call in any information leading to the suspicion of arson. Investigators are given intensive training in the detection of arson, and the courts have given heavy sentences to those convicted. These measures have yielded tangible results: the incidence of intentionally set fires was cut in half in the first two years of the program.

Still, the fight is not easy. Arson is hard to prove because the criminal doesn't carry away loot which could be used as evidence. Most clues, moreover, are burned away in the fire, and unless the accused admits guilt, it is difficult to convince a judge or jury that there is no reasonable doubt of guilt.

There are cases on record where professional arsonists, known as *torches* in the underworld, have actually been burned in the act, but who nevertheless escaped conviction for lack of admissible evidence. A torch may earn as much as $3,500 for a night's work, although six firemen lost their lives in a supermarket fire in the summer of 1978 that was allegedly set by three men who were paid only $500 each.

Not too long ago, only 10 percent of all arsonists were arrested; few of these went to trial, and hardly any of them ever went to jail. This made arson an attractive profession for anyone with a criminal turn of mind. The scene is changing now, and more arsonists are being apprehended, brought to trial, and convicted.

Who are the arsonists, and what makes them what

FIRE! ITS MANY FACES AND MOODS

they are? There is no one answer to this question. Personalities and motives vary, the only common denominator being a hard heart. Like most crimes, arson is a crime without pity.

First on the list is the *pyromaniac*, a mentally unbalanced person who gets a thrill from watching fires. When the origin of a blaze looks suspicious, police scan the crowd of spectators on the chance of spotting someone wearing a look of rapture. Such emotionally distrubed people do not, however, make a large contribution to arson statistics on a national level. Their terror is felt more on a local level when police and fire officials are sometimes driven mad by a rash of set fires. When such an individual is apprehended, he or she usually ends up in a mental hospital rather than in jail.

A second motivation for arson might surprise you: vanity. A certain type of person, in the hope of becoming a hero, lights a fire, waits for the engines to arrive, and then lends a hand in rescuing victims and helping to subduc the flames. More than one such person has been awarded a medal for bravery, only to be arrested at a later date.

The most common motive for arson is revenge, the desire to get even with someone. In New York's South Bronx, twenty-five persons perished recently in a fire at a social club. The fire was set by a man who allegedly was jealous because his girl friend was attending a party against his wishes. In another case a number of people were injured when two fires broke out on

Crime Without Pity

the same day in two New York hotels. A few days later police arrested a man who had been fired by both hotels.

A relatively new motive for arson is an unfortunate by-product of our social welfare system. Welfare recipients living in slum housing have been known to set their own homes on fire in order to qualify for new furniture and relocation allowances.

Sometimes a fire is set to cover up a crime. Whether it be murder or robbery, scorching flames can do a good job of erasing clues. For this reason autopsies are often ordered to see if a charred body contains a bullet or two. In one out of twenty murders, fire is the actual murder weapon.

The most frightening arsonists, however, are the ones who appear most respectable on the surface. They worship in churches and temples, provide for their families, support charities, and patronize the arts. They are the increasing number of businessmen who burn down their own plants to collect insurance or escape financial difficulty.

Police investigators have always known of the existence of this kind of white-collar criminal, but now their ranks have grown to the point that they are responsible for one-third of all deliberately set fires. This mounting rate has resulted in the listing of arson as a Class I crime by the FBI. This places it in the same category as murder, rape, and grand larceny.

Some schemes to collect fire insurance even manage to stay within the letter of the law. A large Eastern

FIRE! ITS MANY FACES AND MOODS

city group recently worked out the following scheme. They bought abandoned tenements, resold them among themselves several times, each time increasing the sale price. These transactions were all on paper; the only money involved was that needed for recording the deeds. With each sale, however, the insurance was increased to keep pace with the growing selling price. When the coverage was high enough, it was time for a fire. It wasn't even necessary to hire a torch: the group simply left the doors of the buildings open. It was just a matter of time before kids would set the places on fire for kicks, or petty thieves would burn it down to get the pipes and plumbing fixtures in order to resell them.

The American Insurance Association is hoping to put an end to such practices by establishing the Property Insurance Loss Register and enlisting the aid of a computer. Claims adjusters throughout the country submit reports on all fire losses of $500 or more. By feeding the details of these fires and claims into the computer's memory bank, it will be easy to pinpoint repetitious circumstances, frequent claims, duplicate policy frauds, and fire patterns suggesting arson.

Another plan to thwart arson for profit would require that all money from insurance claims be used to rebuild the premises rather than being pocketed. At least one major insurer already refuses to pay full replacement value unless replacement actually takes place. Generally, payment is based on the depreciated value of the property.

Crime Without Pity

Insurance companies cannot prosecute suspected arsonists, but they can do the next best thing by refusing to pay the claim. This requires the offender to take the insurer to court, a risky move that often uncovers evidence which then sets the stage for criminal prosecution.

Although most large cities now have arson task forces, the sad fact is that our slumping economy has drained away much of the money needed to keep such agencies effective. In New York City, for example, several hundred fire marshals could be kept busy doing their job. Only recently, however, was their number raised—from 77 to a mere 152. It doesn't take much arithmetic to conclude that the marshals are overworked, and that they must ignore the majority of suspicious blazes. Unless arson is clearly indicated, they don't have time to respond to a fire unless it is a four-alarmer.

Fire marshals usually are men who have served as firemen. Sometimes they are forced out of the front line of duty by injuries. Their job, however, is anything but an act of retirement.

A fire marshal is a detective in every sense of the word. He carries a gun and has the authority to search and arrest. Sometimes he collects evidence by posing as a member of the underworld. By playing such a role, or merely by walking into an empty building for a routine inspection, he puts his life on the line.

Although fire destroys evidence, a fire marshal can spot clues an untrained person would miss. If an en-

FIRE! ITS MANY FACES AND MOODS

tire building goes up in flames at once, or if fires break out simultaneously in different locations in a factory, it is probable that an arsonist has been at work. Trailers, gasoline-soaked material set out in a fuselike path, may have been used to take the fire from the primary point of ignition to other parts of the building. A blocked doorway might indicate an attempt to hamper fire fighters during the vital early stage of the fire. Perhaps the sprinkling system has been tampered with.

The color of the flames tells a story too. The color of a flame depends on the temperature of the fire. If high burning temperatures are indicated in a building that houses only low-temperature materials, something is wrong. Black smoke suggests burning petroleum, whereas nitrocellulose fiber produces yellow smoke. Both are used by arsonists.

If water spreads the flames, the fire is being fed by a flammable liquid. Small licks of flames in puddles of water tell the same story. Explosions, no matter how slight, demand an explanation, and in the absence of a logical solution, arson is the probable answer.

Once the fire is out the fire marshal seeks further clues. Floorboards that are charred inside the cracks show that a flammable liquid seeped in between the boards, indicating that the area was doused with gasoline, naphtha, or another flammable liquid. Burned-out matches can be identified if they are found before being stepped on or swept away. Samples of ashes are collected for laboratory testing. One test looks for hydrocarbons that result from burned petroleum. No lab

Crime Without Pity

equipment, however, can take the place of a keen eye, a good sense of smell, and the intuition developed through experience.

Fire marshals and arson task forces alone cannot win the battle against arson unless ordinary citizens also become involved. In 1977 the city of Boston broke up the largest known arson ring in the United States because the citizens of a burned-out neighborhood decided to fight their own battle. They collected evidence and took it to the state attorney general. Twenty-two persons, including former high-ranking police and fire officials, lawyers, realtors, and insurance brokers were arrested. Later arrests brought the total to thirty-three. This was made possible through the efforts of people just like you.

What can you do right now? First, report any suspicious persons or activity to the police department, the fire department, or the fire marshal. Second, support every community effort to fight arson, the crime that shows contempt for both property and life.

As you will see in the next chapter, fire fighting used to involve throwing buckets of water on the flames. Improvements in equipment have made it possible to subdue fires much more effectively today.

9.
FIGHTING FIRE

Fire!

No cry of alarm is more exciting than this dramatic call. Immediately people band together to subdue the ageless enemy and to rescue victims from its sweep. Ordinary people are moved to acts of extraordinary heroism, and fire fighters risk their lives with no thought of glory or reward, but because it is something that has to be done.

In the past the only response to fire was the thought of escape. As is still the case with forest animals today, life depended on leaping faster than the flames. But when humans began to live in shelters that they built with their own hands, they learned to fight back for what was theirs. Throughout the ages, the struggle has been essentially the same. Modern equipment and technology notwithstanding, the challenge in the majority of fires today is the same as it was when Rome burned in A.D. 64: find water and drown the flames.

At the scene of a fire today, spectators are kept out

Fighting Fire

of harm's way. If you were living in colonial America, you wouldn't be allowed to be a mere spectator. Every householder was required by law to own a leather fire bucket and to keep it filled with water at all times. At night three buckets had to be kept on the front step. At the sound of an alarm, day or night, everyone grabbed a bucket and ran to join the fight.

There were no hydrants. Water was brought to the scene by the formation of a bucket brigade, a double line of people, which linked the fire to the nearest well or stream. The strong side of the line was made up of the most able bodied, who had the strength to pass the brimful containers, while the weaker line was made up of older persons and children, who passed back the empties for refilling. At the head of the line were the strongest men who could hurl the water the greatest distance. Even though piston pumps had been used in ancient Egypt, fire fighting in America was primitive.

The greatest threat of fire to farms and villages came from the combination of fireplace chimneys and thatched roofs. Thatching consisted of tied bundles of dry straw, and when a spark landed on such a roof, fire was almost a certainty. If the first plumes of smoke were detected, a single splash of water was enough to put it out. If a fire got out of control, however, it was necessary to use the *hook*. A large iron hook attached to the end of a long rope would be thrown to the roof, and with the strength of several men the roof would be torn loose and pulled to the ground. Thatched roofs

FIRE! ITS MANY FACES AND MOODS

are long gone, but hook-and-ladder companies are still familiar in fire departments today.

The ladder is something you may take for granted. When firemen respond to city fires today, tall aerial ladders reach for the roof the moment the engines arrive. In colonial days it was not quite as simple. When fire broke out in the city, people ran to a central location where the ladders were stored and then dashed with them through the streets, often bowling over everyone who got in their way.

As time passed and buildings grew taller, hand ladders were no longer adequate. As late as 1871, most of the business district losses in the Chicago Fire resulted from the lack of ladders long enough to reach the upper floors. This experience hastened the invention of the aerial ladder, which soon became available to city fire departments.

The age of the bucket brigade began to fade with the invention of the hand pumper. It is said that the first such fire engine in America was delivered from England to the city of Boston in 1679. Now, at the sound of an alarm, a dozen or more men grabbed long dragropes and hauled the heavy wagon through the streets at breakneck speed. That was an act of courage in itself, because if one of them stumbled and fell, he would almost certainly be run over by the heavy apparatus. At night, a lad with a lantern would run ahead of the wagon to light the way and clear the path.

Fighting Fire

Each hand pumper carried a supply of water, and often this supply was enough to put out the blaze. At bigger fires, however, a bucket brigade still had to be formed to keep the square wooden tub filled. Bucket brigades disappeared altogether when suction hoses were designed to get water to the pumper from a nearby well or stream. If the source of water was some distance away, a number of pumpers would link together, each supplying water to the next, until the last wagon was able to play a stream of water on the fire.

When such a tandem arrangement became necessary, the volunteer crews working each rig fought not only to put out the fire, but also to enhance their reputation. If a rival company filled the tub faster than they themselves could empty it, the tank would overflow. This was called being *washed*, and it was considered a terrible disgrace.

Operating the hand pumpers was no easy task. Maintaining a pace of 60 to 120 strokes per minute, the men working the long wooden handles would exhaust themselves and would have to be relieved every ten minutes. Consequently, a crew of 50 to 80 volunteers was needed to operate a single unit.

Early hand pumpers were equipped with stationary water cannons mounted on the wagon. This changed when someone learned how to make hoses out of ox entrails. These were later replaced with hoses of leather, held together by stitching or riveting, and the

FIRE! ITS MANY FACES AND MOODS

rubber hose was invented in 1871. Today, rubber hoses are covered with strong material to protect them from rough wear.

We don't think of different fire companies as rivals anymore, except perhaps during competitions at county fairs. Among early nineteenth century fire companies, however, the rivalry was often more important than putting out the fire. In the race to get to the fire first, any kind of dirty trick was allowed. Fires often burned briskly while firemen engaged in street brawls to determine which company had the right-of-way down a narrow street or who should get to use the nearest hydrant. Hydrants were sometimes hidden under an empty barrel to keep competing companies from finding them. Confusion was magnified by the common practice of breaking open a cask of rum to keep spirits high while battling a blaze.

Excessive competition was also encouraged by the insurance companies' policy of paying a bonus to the brigade that arrived first to put out a fire. The reward was paid by the company that insured that particular building. Most buildings were identified with *fire marks*, emblems posted on outside walls to show which particular agencies protected them. This grew from the practice in England whereby the different insurance companies employed their own fire brigades. If a burning building wasn't posted with the correct fire mark, the fire fighters would return to their station without even attempting to put it out.

Fighting Fire

The invention of the steamer greatly increased the capability of fire fighters. One steam engine could do the work of seventy-five firemen and twelve hand pumpers. Pressure from the steam boiler could push water through 350 feet of hose, still leaving enough nozzle pressure to shoot a stream of water a distance of 130 feet. But the fire fighters weren't too happy because the large boilers were too heavy to haul.

In 1850 a rowdy group of Cincinnati volunteers refused to work a steam pumper the city had recently purchased. When a large fire broke out, citizens took matters into their own hands and fought a pitched battle with the defiant rebels. The volunteers were driven off with rocks and clubs, whereupon the citizens manned the steamer and put out the blaze.

New York City got its first steam engine in 1865, the same year the Civil War ended. When someone suggested that horses be used to pull the heavy wagons, fire fighters protested again. They didn't want to turn their firehouses into stables: who wants to clean up after horses? Practicality won out, however, and it wasn't long before firemen learned to like their horses as much as they liked each other.

Firehorses were remarkably well trained. No one knew this better than a milkman who bought a retired firehorse to pull his wagon, only to end up being dragged to every fire thereafter. When the firehouse alarm rang, the horses, prodded with the flick of a mechanical whip, trotted out and positioned them-

FIRE! ITS MANY FACES AND MOODS

selves in front of the steam engine. The harness hung from the ceiling, and when the driver tugged the reins it dropped onto the horses and was secured in less than a minute. Meanwhile, firemen were sliding down the brass pole from their bunkroom on the second floor. The brass pole was introduced in the 1870s, and it is still common in firehouses today.

Fire engines were now too fast to have a boy run ahead of them, so Dalmatians were trained to take over the responsibility of clearing the way. Firehouse dogs of all breeds have won many medals for valor, and legends about them are plentiful enough to fill a whole separate book.

The sight of three charging horses and their steamer tearing along cobblestone streets is an excitement that disappeared with the invention of the gasoline engine. Even today, however, how many of us can resist the urge to turn and watch each time a fire engine barrels past?

In Roman times firemen used a brass syringe that could squirt a few quarts of water at a time. You would consider it little more than a large water-pistol. It was 3 feet long and had to be worked by three men: two men held it while a third worked the plunger, pulling it back to draw water from the tub and pushing it forward to shoot water onto the fire. The slow pace of evolution is demonstrated by the fact that similar instruments were still being used in the 1700s.

Fighting Fire

After the fall of the Roman Empire, however, Europe forgot about fire fighting. These were the Dark Ages, and religious leaders taught that fire was the just wrath of God. Preachers proclaimed that any attempt to put out a fire was a sacrilegious act in defiance of God's will. During the sixteenth century a revival of interest in controlling fires resulted from the translation of ancient Greek manuscripts describing the design of fire-fighting apparatus. It wasn't until after the Fire of London, in 1666, however, that serious efforts were made to resume the development of fire-fighting equipment and skills.

A serious challenge to modern fire departments is the spread of high-rise buildings. At one time, tall buildings were erected only in large cities, and city fire departments have long had the know-how to handle upper-story fires. Today, however, many small suburban communities with limited equipment are facing the challenge attendant on the increasing construction of skyscraper motels, condominiums, and office buildings.

When a building has more than seven floors, special precautions have to be taken. The most obvious is the construction of roof tanks which contain enough water to service the building as well as to put out a fire. Also necessary are standpipe connections at sidewalk level. Pumpers force water up these pipes with sufficient pressure to reach the upper floors. If a building is very high, relay pumping stations at several levels

FIRE! ITS MANY FACES AND MOODS

insure adequate pressure all the way to the top. Some standpipes lead to hose outlets, whereas others are connected to sprinkler systems.

The importance of roof tanks was illustrated dramatically in New York City way back in April, 1927. The Sherry Netherlands Hotel was under construction, and its unfinished upper floors were encased in wooden scaffolding used by bricklayers and other construction workers. One day about noontime someone looked up from the street and gasped in disbelief. Soon all eyes were turned upward. At first all they could see was smoke, but soon leaping flames began to eat away at the scaffolding. Fire engines arrived in a matter of minutes, but the roof tanks were not yet operational. Like the crowd of spectators, the firemen could do nothing but stand and watch as the top of the skyscraper blazed like a torch.

If only the firemen then had had a new piece of equipment that is being developed today by the McDonnell Douglas Corporation. The *Suspended Maneuvering System* (SMS) is a self-controlled 1-ton module cage that can be carried by helicopter to the scene of a skyscraper fire. Hanging from a steel cable 1,000 feet long, it will be able to roam 150 feet in front or in back of the helicopter and dock on the roof or at a window ledge of a burning building. A 7-by-8-foot platform is capable of carrying eight fully-equipped firemen and eighteen persons. Its usefulness isn't restricted to fire operations. It can go to the aid of people trapped by floods, earthquake, forest fires,

Fighting Fire

on burning ships at sea. Despite an anticipated high purchase price, probably most large cities will invest in a device of this type; since helicopters have a wide flying range, suburban communities will benefit as well. At present there are more than four thousand skyscraper fires a year in the United States, and this number will continue to increase. The need for equipment such as the SMS is clear.

The next time you're watching firemen battle a blaze you might enjoy recognizing the different types of apparatus and knowing each one's function.

Pumpers are recognized by their heavy, squat build and the many brass or chrome fittings and dials built into their sides. They are used to supply hoses with water at a satisfactory pressure. A hose delivery rate of 325 gallons per minute is common, compared to the 5 gallons per minute that you get from a garden hose. If you have ever seen a garden hose whip around under pressure, you'll understand why it takes three firemen to man a single line. Most pumpers are triple combinations, containing fire pumps, hose compartments, and water tanks.

To keep the water tank up to a satisfactory level, the pump man connects the pump to a hydrant near the scene of the fire. The pump man must be a competent engineer: to deliver water at the right pressure he must calculate the length of the hoses, the number being used, the amount of water friction, the size of the nozzles, and hydrant pressure. If fire climbs to

FIRE! ITS MANY FACES AND MOODS

the upper floors of a building, the pump operator has to increase pressure so that the water will reach the upper windows. Pumpers usually carry an extension ladder, a roof ladder, and other tools and equipment such as fire axes and hand extinguishers.

New York City has a superpumper that connects to three satellite pumpers, all purchased for a total of almost a million dollars. It is so big that it has to be fed from three hydrants or from the enormous draft of a fire boat. It can supply nine 4½-inch hoses or thirty-five smaller lines. The pumper and each satellite have deck water cannons that can knock down brick walls.

Aerial ladder trucks, sometimes called "big sticks," are either double-axled or trailer-drawn. In addition to the 60- to 100-foot mechanically operated ladder, each truck carries about 200 feet of manually operated ground ladders. Most have water pipes for training heavy streams of water at upper levels. Early aerial ladders were cranked into position by hand, but today hydraulic power is used and the operator on ground level needs only to manipulate a set of buttons.

Sometimes aerial ladders are mounted on pumpers, making for quadruple combinations. Aerial companies also carry equipment and tools for entry, ventilation, salvage, and rescue. This equipment is mounted on the back of tractor-drawn units; these are the ones where you see a second driver who turns the rear wheels independently of the tractor wheels. Otherwise, the long apparatus would never be able to turn

Fighting Fire

a sharp corner. Probably every city lad has dreamt at least once of being a tillerman on the tail end of a "hook-n-ladder."

Aerial platforms, often called "snorkels," are king-sized brothers of the equipment used by telephone repairmen to reach wires atop telephone poles. In position, it has the appearance of a huge jointed cobra ready to strike. The platforms provide an elevated base from which to fight a blaze, as well as an escape chamber for victims trapped on upper stories. From this vantage point a fire captain is in a good position to coordinate and direct operations. Most platforms reach a height of about 90 feet.

Water towers have the same design as aerial platforms, but they are not manned at the top. Controls on the truck aim the nozzle, determine the size of the spray, and control the water flow.

These are the most common apparatuses seen at the scene of a fire. The functions of other pieces of equipment are usually easy to spot. Tank apparatuses bring water to locations where hydrants are not available. Ambulances, heavy and light rescue trucks, searchlight units, and smoke ejectors all play a role in fire fighting and rescue operations.

At one time it was common to see firemen clinging to the side and back platforms of a fire apparatus flying along in response to an alarm. Their only support was brass handrails. More than one fire fighter has been thrown from a truck and killed. Modern equipment eliminates this needless hazard. The next time

FIRE! ITS MANY FACES AND MOODS

you see an engine go by, note that each member of the entire company is safely buckled into his seat.

Every city has its own particular fire problems. Waterfronts must be protected by fireboats. No other equipment can match their pumping capability. New York City's fireboats are capable of shooting water 225 feet upwards to the road levels of the George Washington Bridge. This is quite a bit different from the situation in 1809, when the city bought its first fireboat, one that had to be hand-rowed and hand-pumped. The newest thing in fighting pier fires is teams of scuba divers who swim among the pilings and shoot water upwards from floating sleds.

In addition to getting help from the local city fire department, airports maintain their own fire and rescue equipment. At a moment's warning they can blanket a runway with foam in the event of a crash landing.

Even New York City's Empire State Building has its own fire department, and rarely does it have to call on the City for help. On one notable occasion, however, the combined efficiency of both departments prevented what might have been a major disaster. In 1943 an Army B-25 bomber crashed into the seventy-eighth and seventy-ninth floors, tearing a whopping 18-by-20-foot hole in the side of the majestic landmark. Flaming fuel and wreckage gushed across the floor until it came out the opposite side of the build-

Fighting Fire

ing, setting the penthouse of a neighboring structure aflame. Fire tumbled down an elevator shaft.

The three crew members of the plane and eleven women who worked in the stricken offices were killed. Considering the disaster potential, however, the death toll was low. The safety design of the building and the sharp efficiency and cooperation of private and municipal fire fighters saved the day.

The key to successful fire fighting is quick response. In ancient Rome, slaves patrolled the walls of the city at night so as to raise the alarm should fire break out. The American colonies had "rattle watches," men on fire watch who carried wooden clackers similar to the noisemakers used on New Year's Eve. And for many generations the peal of the village fire bell was a sound that could be heard for miles around.

The nineteenth century brought electronic fire alarms, the familiar red boxes you see attached to street poles and the sides of buildings. If you live in a smaller city, coded gongs ring in every firehouse the moment the lever is pulled. Only fire companies close to the box respond, unless additional alarms are registered. In large cities the signal goes to a fire communication center, from which instructions are relayed to companies in strategic locations.

One problem with electronic alarms is the ease with which false alarms can be sounded. Telephone alarm boxes are now emerging as the most effective

FIRE! ITS MANY FACES AND MOODS

way to eliminate false alarms: people will think twice before making a false alarm if they know their voice is being recorded. Also, a person who is sounding a valid alarm will have the chance to describe the fire or the nature of the emergency. This will enable the dispatcher to send out whatever equipment is most appropriate.

The most interesting way to learn about fire fighting is to go to one of the many museums maintained by fire departments, or, better still, to visit an actual firehouse. Firemen take pride in themselves and in their equipment, and they love to show visitors around. Maybe you could arrange for a fire fighting demonstration at your school. Now let's look at the fire fighters themselves and see what makes them such special people.

10.
FIRE FIGHTERS

Who are the people you see riding the fire engines? What kind of people climb ladders coated with ice and walk on roofs that could cave in at any moment and plunge them to a fiery death?

There are close to one-and-a-half million fire fighters, men and women, in the United States. Each year half of them are injured. Every three days one of them is killed, a death rate five times greater than that of policemen. What makes them accept such fearful odds? Are they supermen and -women?

The answer is yes, but not like Clark Kent. As a New York fire chief once put it, the fire fighter performs an act of bravery the day he or she joins the department. From then on it is just a matter of doing a job and not thinking about getting killed. The goal of a fire fighter is to put out fires and saves lives, not to be a hero.

Contrary to what you might think, most firemen who die in action do not perish from flames or smoke.

FIRE! ITS MANY FACES AND MOODS

Heart attacks caused by stress account for more than 40 percent of line-of-duty fire-fighter fatalities. This compares to 25 percent from wounds and loss of blood, 10 percent from smoke inhalation, 9 percent from burns, and the remainder from drowning and other causes.

If you think about it, the heart attack death rate is understandable. Most of us rise for work in a leisurely fashion. We set the alarm for another ten minutes. We enjoy a refreshing shower and dress at a slow pace. We sit down and enjoy breakfast before leaving the house.

In sharp contrast, the fireman never knows when the alarm will sound. If he is sleeping he must leap from bed in an instant, get dressed, and within minutes he is clinging for life to a fire engine as it careens through streets in all kinds of weather. Arriving at the fire, there is the hard routine of dragging hose lines, setting and climbing ladders, as well as chopping open passageways and struggling through smoke-filled rooms in search of victims. All of this takes place in less time than it takes some of us to eat breakfast. Is it any wonder that many firemen buckle under the strain?

When a rookie is assigned to a company, he is given a chance to do a little bit of everything. After a time, depending on his particular talents and the needs of the department, he is given a specific assignment.

Laddermen advance hose lines, vent roofs, and

Fire Fighters

make rescues. Engineers work the pumpers. Hosemen fight the blaze and wet down adjacent areas to prevent the spread of the fire. Salvage men protect property both during and after the fire. Water damage often exceeds damage from flame and smoke, and an efficient salvage crew can prevent millions of dollars in property loss by spreading tarpaulins and channeling water out of a burning building. Drivers have the job of jockeying heavy apparatus through traffic, maneuvering it close to hydrants, and then joining others where he is needed. Each person in the company has a job to do; nothing is left to chance. The order of priority in fire fighting is rescue, containment, and attack.

The role of a modern fire department goes far beyond the obvious responsibility of putting out fires, and the job of the individual fireman doesn't end with climbing ladders and manning water cannons. Lifesaving and paramedical services are a part of every fire fighter's duty. Fire prevention is even more important than fire fighting, and firemen regularly inspect factories and businesses in their district. Many fires are able to spread as a result of uncharged fire extinguishers, blocked passageways, locked sprinkler valves, accumulated refuse, and poor maintenance.

Safety education isn't restricted to Fire Prevention Week (which dates back to 1911 when it was designated to take place during the week of October 9, the anniversary of the Great Chicago Fire of 1871). Public relations and community education are year-

FIRE! ITS MANY FACES AND MOODS

round assignments. Add to the above-named activities data development, regional coordination, and developmental planning, and you begin to understand that firemen don't just sit around playing checkers or watching TV and waiting for bells to ring.

In a city, the departmental chain of command begins with a fire commissioner and a fire chief. Deputy chiefs command battalions, each of which is made up of four or more companies. Each company is commanded by a captain aided by company lieutenants, who pass down commands to the men. Top officials usually show up only at major fires, which sometimes even bring out the mayor. Older New Yorkers will always cherish the memory of Mayor Fiorello LaGuardia (in office 1934–45), an uncontrollable fire buff who would dash from his desk at the sound of sirens, don a helmet, and leap aboard the tail platform of an engine responding to an alarm.

The military structure of a fire department largely explains its efficient operation as well as the immaculate maintenance of firehouse and equipment. It is a tradition to keep engines and equipment in showroom condition, and after each fire many hours are spent drying out hoses, shining brass, and cleaning up the mess that always attends fires. Ladder drills and other exercises are frequent, and blackboard instruction gets top priority. An individual class session might explain the use of a new piece of equipment or study in detail the layout of a plant located in the district. Such devotion to advance preparation is evident at the Massachusetts Firefighting Academy, where an operational model of a burning liquified-gas tank was recently constructed at a cost of $250,000. When liquified gas goes on fire it explodes the tank, sending a murderous fireball into the sky. With training in how to handle such fires, Massachusetts fire fighters stand a better chance of doing their job and surviving.

When a fireman gives his life in the line of duty, it strikes the hearts of fire fighters everywhere. Uniformed guards of honor often travel from distant states to pay their respects to a fallen comrade. The question in everyone's mind is always the same. Did it have to happen?

The risks will never cease to exist, but much is being done to reduce them. The traditional fireman's helmet, romantic though it may be, will someday be replaced by one which is superior in both material and design. Most are made of hard plastic, not the best thing to wear in the superheated air of a bad fire. The present design is such that the helmets snag with breathing apparatus. In the future the helmet and face mask will be an integrated unit. There is room for improvement in materials as well: the present standard American helmet can withstand a blow of only 40 pounds, one-third the strength of the British helmet.

FIRE! ITS MANY FACES AND MOODS

Today's breathing apparatus lasts only thirty minutes on one tank of air, and it weighs 30 pounds, which is much too heavy. Improvement is called for, particularly since toxic injuries are causing some cities to make the use of masks standard procedure at all fires.

Even the fireman's turnout coat could stand improvement. Not only does it hinder movement because of its bulk, but it also becomes a sweatbox when the man is in action. In addition, it is not constructed to resist blows, a serious drawback, since injuries are twenty-six times more common than burns.

A great deal has been done already to reduce the risk of heart strain. Central communications headquarters now spare the men in each firehouse the tension of decoding the gongs each time a fire is rung in from somewhere in the city. The men hear only the particular signals to which they must respond.

Much improvement has also been made in the work-shift scheduling of firemen. In the past, fire fighters rarely saw their wives and children. Now, every time a firemen has to put in a long stretch, he is compensated with a few days of unbroken rest.

One problem, however, needs to be addressed. A fireman's pay is low, particularly considering the risks he faces. Many men, particularly young men raising a family, are forced by finances to hold down a second job. Fire fighting is a very demanding full-time job, and the men should enjoy economic security without

Fire Fighters

having to take on additional burdens. The resolution of this problem can come only through the support of an educated community.

Do you think you would like to be a fire fighter? If so, you'll have to plan your moves very carefully because the competition is tough and it takes a lot to qualify.

First of all, don't quit school. The days of brawn ended a long time ago. Fire fighters today are highly trained engineers, and you have to pass a difficult qualifying examination to be considered for an appointment. Even volunteer companies insist that rookies undergo intensive training, both in the field and in the classroom.

You must be certain that you are mentally suited for the job. No department wants an immature glory hound as a rookie. Are you ready to polish brass and shovel snow? Are you willing to work on Christmas Day?

You should read a few books written by and about firemen to get a clearer understanding of what the job really involves. Visit a local firehouse and talk to some of the men. Listen to their stories and learn from their experiences.

In order to take the Civil Service examination required of all candidates you must have at least a high school diploma. Many departments require at least two years of college with a major in fire science. Even if you score high on the exam, you won't be appointed unless you pass several demanding physical and psychological examinations. Acrophobia, the fear of heights, would rule you out immediately. You might not be emotionally suited to the militaristic operational routine of a fire department. You have to know how to take orders.

If you live in a small community you might want to become a volunteer fireman. The fact that the work is unpaid doesn't necessarily make it easy to be accepted, because the waiting list is almost always long. Volunteers make a considerable sacrifice because they serve in their free time after completing an ordinary day's work elsewhere. People who work locally and can walk away from their job when the alarm sounds make the best volunteer candidates.

If you are accepted, whether as a volunteer or for pay, don't expect to be out fighting fires the following day. You will go through three or more months of training at a fire college where you'll work as hard as a Marine going through basic training.

You will learn how to carry a man on your shoulders as you climb down a 100-foot ladder; how to rappel down a rope from the roof of a six-story house; how to set a broken arm; how to get a cat out of a tree; and how to hold onto a ladder with your legs while working a hose with your hands.

The aim of this training is not to become a chief, but just to have the opportunity of being accepted as a rookie. If this is what you want, don't be discouraged. The reward of being a fireman is more than worth the effort.

Many aspects of fire and fire fighting have now been covered. The next chapter explains what fire actually is.

11.
WHAT IS FIRE?

Up to this point you have read a lot about fire, but the most basic of all questions still remains to be answered. What is fire?

The answer to this question is not as simple as you might think. Recognizing fire is one thing; explaining it is something else—rather like trying to describe a corkscrew without using your hands.

If you had gone to school in the eighteenth century, you would have been taught that something will burn if it contains *phlogiston*, the principle of fire, and that what we call fire is simply the escape of this substance.

This was a neat explanation, except for one problem. It would follow that when something lost phlogiston as a result of burning, it would end up weighing less. A scientist named Georg Stahl, however, demonstrated that the total remains of a fire and all of its by-products exceed the original weight of the object consumed. This was an astounding discovery.

What is Fire?

How can something get bigger by being destroyed? Naturally, the phlogiston theory had to be scrapped.

To know what fire really is, you must understand two things: the nature and behavior of molecules and the personality of oxygen.

You have probably learned in basic science that all things are composed of tiny, invisible elements called atoms. Although a drop of water contains more atoms than there are people living on the earth, there are only 105 *kinds* of atoms. We know some of them, such as hydrogen and oxygen, as gases and others, such as gold and silver, as solids. In our normal earth environment, we encounter only two elements as liquids—bromine and mercury. If you are curious about the rest, you can find a complete table of the elements in any chemistry textbook.

Atoms combine in different patterns to form molecules, and this accounts for the existence of different materials. A drop of water is different from a pebble because they have different molecular structures.

Molecules are always in motion, though of course you can't see this motion with your eyes. Whether something is a gas, a liquid, or a solid depends on the rapidity of its molecular activity. Motion is slow in a solid, faster in a liquid, and fastest in a gas. In addition, it's important to know that heat produces an increase in molecular activity. This is why ice turns into water and water changes to steam as the temperature rises.

FIRE! ITS MANY FACES AND MOODS

Oxygen was discovered in 1774 by Joseph Priestly. Soon afterwards it was determined that one-fifth of the air that surrounds the earth is made up of oxygen.

What does all this have to do with fire? You'll get the answer by reading just a bit further.

The Fire Triangle

The *fire triangle* is made up of three essential parts: fuel, heat, and oxygen. If any one of these three parts is missing, there can be no fire. At the same time, the three will produce fire only if they unite in a satisfactory combination. Let's look at each in turn to see how everything fits together.

Fuel

Since you were a child you've known that some things burn and others don't. But have you ever wondered why? Gasoline is a liquid and it burns like fury, whereas water, also a liquid, puts fires out. Coal, a solid, is one of the world's primary fuels, but you know that the rocks in your backyard won't burn at all. Is it any wonder that early scientists figured that some things have phlogiston and others do not?

The truth lies in the story of atoms and molecules. Of the 105 elements, only particular ones are capable of burning. Two of these, *hydrogen* and *carbon*, account for the vast majority of the fires that you see and read about.

Wood burns because it contains *cellulose*, and cellulose is made up of six atoms of carbon, ten atoms of

What is Fire?

hydrogen, and five atoms of oxygen. A chemist would write it this way: $C_6H_{10}O_5$

Gasoline burns violently because, like all petroleum fuels, it is made up entirely of hydrogen and carbon. For this reason petroleum products are called hydrocarbons.

Heat

Just because something *can* burn doesn't mean that it *will*. A log will sit in a fireplace forever unless you introduce the second side of the triangle, heat.

Heat is harder to explain than it is to feel. Your knowledge of molecules, however, can help you to understand why you feel heat, which is nothing more than rapid molecular activity. When you rub your hands together briskly, your skin molecules vibrate and your hands feel warm. You don't have enough strength to snap a wire clothes hanger, but if you keep bending it back and forth at the same point, it will get hot and snap.

Heat can be produced with mechanical, electrical, chemical, or nuclear energy. The details are too complicated for our present purpose. The important point is that no fuel will burn unless it is heated, regardless of the heat.

Oxygen

The final element in the fire triangle is oxygen. In almost all fires the oxygen is taken from the air, which is one-fifth oxygen.

FIRE! ITS MANY FACES AND MOODS

When a fire's oxygen supply is increased, it will intensify. This is why you blow air into a fire to encourage the flames. It also explains how to put out a fire in a frying pan. Simply cut off the oxygen supply by placing a lid on top.

Oxygen doesn't burn by itself, but is necessary for combustion.

Now that you know the three parts of the fire triangle, it remains only to put them together to understand the anatomy of a fire.

You start with fuel, something that will burn. Almost always this is something whose molecules include hydrogen and carbon atoms. Then you must raise the temperature of that substance by applying heat. This causes molecules to vibrate and, unseen by the eye, the molecules break down and atoms begin to escape. They unite with the oxygen in the air, a process called *oxidation*.

Two points should be noted. First, not all heated objects will oxidize when exposed to the air. Oxidation takes place only if the surface atoms can combine with oxygen. Second, the oxidation process can be extremely slow. The rusting of a nail is an example of slow oxidation. It can be said that the nail is burning at a very slow rate.

When the rate of oxidation reaches a certain point, radiant energy and heat are released as by-products. This is fire.

Fire, then, is rapid oxidation, and once it begins

What is Fire?

it feeds upon itself. Heat and flames reach for new fuel, creating a sustained chain reaction.

With the knowledge of what fire really is, you will now understand the solution to the mystery mentioned earlier in the chapter. How can the total by-products of a fire weigh more than the object before it was burned? The answer is that part of the fire, oxygen, is gathered from the air and added to the original material.

Now that we know what fire is, let's examine its characteristics—the personality of fire.

12.
THE PERSONALITY OF FIRE

Like every human being, you have a personality. It is made up of your own particular characteristics, your manner, your way of behaving. Your personality is what makes you different from everyone else.

As a living force, fire too has a personality, and the better you understand its traits, the better you will comprehend its behavior. We should look into some of the reasons why it acts the way it does.

Heat

As a simple experiment you might take two pots, one containing water and the second an equal amount of oil, and place them over two stove burners that have the same setting. After a time the water will begin to boil vigorously, but the oil will still be placid. Common sense tells you, however, that they both have absorbed the same amount of heat. Indeed, if you put a cooking thermometer into both liquids at the

The Personality of Fire

moment that the water boils, both would read 100° C.

Leave both pots on the burners and the water will begin to boil away into steam and vapor, but the temperature of the water will remain the same. The oil, however, will continue to absorb additional heat because it boils at a much higher temperature. This explains why cooking oil is needed to prepare french fries, and why a burn from boiling oil is much more severe than one from boiling water.

Not all fires produce the same amount of heat. The heat generated by a fire depends on two factors: the nature of the fuel and the amount of oxygen feeding the flames. Not even all kinds of wood burn with the same intensity. If you use a fireplace you'll get much more heat from hickory than you will from pine. It is easier to start a pine fire, however, because pine has a lower *kindling point*, the temperature at which a material breaks into flame.

The amount of heat released by a burning substance is called *heat of combustion*, and it is measured either in *British Thermal Units* (BTUs) or calories (cal). A BTU is the amount of heat required to raise the temperature of one pound of water 1° F. A calorie will raise one gram of water 1° C. If you burn a pound of paper, the fire will release 6,000 BTUs, whereas a pound of coal would give off 12,500 BTUs. A pound of burning gasoline provides 20,000 BTUs. From this you can see why fire fighters like to know what's inside a burning building before they enter it.

FIRE! ITS MANY FACES AND MOODS

Heat and Temperature

Although they are related to each other, heat and temperature are not the same. The body temperature of a baby is the same as that of an adult, but the amount of heat in an infant's body is considerably less. Heat and temperature are directly related only when all other things are equal. If two containers with equal amounts of water are exposed to the same heat, the temperature rise will be identical. A given amount of heat might boil the water in a small vessel, however, while the temperature of the water in a large vessel would remain almost unchanged.

Temperature tells you how fast the molecules of a substance are vibrating. When a mercury thermometer is dipped into hot water, the rapid vibration of the water is picked up by the mercury, causing the mercury to heat and expand, making for a higher reading. Heat measurements, on the other hand, refer to the *entirety* of the molecular disorder in a system. Simply stated, a tub of water contains much more heat than a glass of water of the same temperature. If the total heat in your body were introduced into a small glass of water, the temperature rise would make the water boil.

Heat as perceived by the mind is a relative thing. This is very easy to demonstrate. Set out three bowls of water, cold on the left, room temperature in the middle, and hot on the right. Place your left hand in the cold bowl and your right hand in the hot bowl

The Personality of Fire

and leave them there for a minute. Now remove both hands and place them in the middle bowl. Your left hand will feel warm and your right hand will feel cold, even though they are experiencing the same temperature.

The mind's reaction is relative to the previous condition of each hand. A thermometer wouldn't be tricked in the same way. If you use two thermometers instead of your hands, they will give identical readings when submerged in the middle bowl.

The first thermometer was made by the famous Galileo Galilei in 1592. He took a glass bulb with a long hollow stem and heated it. This caused the air inside the bulb to expand and exit through the stem, creating a partial vacuum in the bulb. He then inserted the stem downward into a container of liquid. As the air in the globe cooled it contracted, drawing some liquid partway up into the tube. Temperature changes caused the air in the bulb to expand or contract, causing the liquid level in the stem to vary. In this instrument if the level went down, an increase in temperature was indicated because the effect followed the expansion of heated air inside the globe.

Galileo's thermometer indicated changes in temperature, but there was no measuring scale. Accuracy was impossible because air was the thermometric (heat measuring) substance, and the volume of the air didn't vary too visibly with heat and cold. In the mid-1600s, more accurate thermometers were designed in which alcohol was the thermometric substance, and finally, in 1714, Gabriel D. Fahrenheit designed an instrument using mercury; this thermometric material is still in use today.

Mercury is very sensitive to heat, and even small changes in temperature will cause it to expand or contract, moving it up and down the hairline tube in the stem of the thermometer. You can't "read" the temperature, however, until you mark a scale on the stem.

A scale of measurement requires reference points, and in dealing with temperature, convenient points of reference are the freezing and boiling points of water. Fahrenheit assigned the number 32 to the freezing temperature of water and the number 212 to the boiling temperature. By dividing the stem distance between these two points, other temperatures can be established. The body temperature of a healthy person, for example, should be 98.2° F.

In America the Fahrenheit scale has been the standard, and you are almost certainly conditioned to thinking in terms of Fahrenheit degrees. If someone says that the outside temperature is 20°, you automatically reach for a coat before leaving the house.

The trouble is that if you were traveling through Europe, a temperature of 20° would be quite mild and you would be quite comfortable in light clothing. This is because in most of the world the *Celsius* scale is used rather than the *Fahrenheit* scale. Now that the United States is in the process of changing over to the metric system of measurement, we all have to

The Personality of Fire

learn to start thinking in terms of Celsius as well as Fahrenheit.

The Celsius scale was established in 1742 by a Swedish astronomer, Anders Celsius. He indicated the numeral 0 as the freezing point of water and 100 as the boiling point. Between these two points the thermometer scale is divided into a hundred equal parts. This explains the often used Latin designation, the *centigrade* (100-degree) scale.

It is very convenient to know how to convert from one system to the other. If you've had some algebra the formulas are easy to use. The first formula is used to change from Fahrenheit to Celsius.

$$C = 5/9 \, (F - 32)$$

In plain English, you subtract 32 from the Fahrenheit reading and multiply the result by 5/9. Suppose the outside temperature is 86° F. Subtract 32 from 86, which gives you 54. Now multiply by 5/9 and you get 30. This tells you that 86° F is the same as 30° C.

To change from Celsius to Fahrenheit you use this equation.

$$F = (9/5 \, C) + 32$$

Take 50° C as the given temperature. First, multiply 50 by 9/5, getting 90. Now add 32 to 90; the result is 112. You have determined that 50° C is equal to 112° F.

It is worth repeating that mercury thermometers work on a very fundamental principle: mercury ex-

FIRE! ITS MANY FACES AND MOODS

pands when it is heated and contracts when it is cooled. Dial thermometers use the same principle. The dial is attached to a hairspring that expands and contracts, moving the dial back and forth.

According to the law of *thermal expansion,* heat causes materials to expand. There is, however, one interesting exception to this law: when water freezes it expands, and this is very fortunate for the human race. By expanding, ice becomes lighter than water and it floats. If surface ice became heavy and sank to the bottom of lakes and rivers, new ice would form on the surface and this too would sink. In a severe winter, lakes and rivers would soon be frozen solid—so solid that no summer would be long enough to melt it. Except for tropical regions, the earth would be a giant icecap!

Heat Transfer

If you've ever touched a hot stove, you know that heat can travel. The molecules of a hot stove are vibrating at a rapid rate, and when you touch the stove your skin molecules vibrate at the same rate. This causes skin damage commonly known as a burn.

This kind of heat transfer is called *conduction,* the process of heating by contact. When you fry bacon and eggs for breakfast, you are cooking by conduction.

If you place your hand alongside the flame of a burning candle, you won't get burned. Place your hand above the flame, however, and it's a different

The Personality of Fire

story. Heat travels in an upward direction, and this process is called *convection*. When you put a marshmallow on the end of a stick and hold it over a fire, you are toasting it by means of convection.

Convection is not hard to understand. The heated air over a flame expands and becomes lighter. This causes it to rise, and heavier cool air rushes in to fill the gap. In turn this cool air is heated, causing it to rise, and the process continues. The rising heat is truly a flowing current, a fact understood by many species of birds that use thermal updrafts to glide.

Because of convection currents, smoke from a fireplace is drawn up the chimney. Unfortunately, much of the heat is lost in the same way, making the fireplace an inefficient means of heating a room. For this reason Benjamin Franklin invented the Franklin Stove, a cast-iron potbellied unit that sits in the center of the room with a horizontal chimney duct to carry away the smoke. Many such stoves are still being used in New England farm houses and general stores, and even city apartments.

Radiation is the third way by which heat is transferred. If you have ever sunbathed you know about radiation. Heat from the sun travels thousands of miles through a vacuum, aided neither by conduction nor convection.

Fire fighters know well the power of radiant heat. It doesn't always take flying sparks to spread a fire. If a building is burning intensely, radiated heat can cause a second building across the street to burst into

FIRE! ITS MANY FACES AND MOODS

flames with no visible sign of ignition. To prevent this from happening, firemen often shoot up curtains of water as a shield against radiation.

Smoke

Most fires produce smoke. Sometimes the smoke is made on purpose, as in the curing of ham and bacon. Most often, however, smoke is an undesirable by-product of fire—unburned fuel. If you see black smoke coming from the chimney of your house, it means that costly fuel is flying away.

When fuel is too rich for the amount of oxygen available, unburned carbon molecules are carried away by convection currents. In your home furnace this can be corrected by thinning the fuel spray of the burner and by cleaning the furnace to allow a greater intake of oxygen.

Carbon monoxide, one of the deadliest of all gases, is another product of incomplete combustion. The most common source of carbon monoxide is the tail pipe of the automobile. Fortunately, this gas is lighter than air and it rises. Other by-products of incomplete gasoline combustion contribute significantly to the blankets of smog that hang over and blight most cities.

Light

When certain kinds of molecules are heated, they give off light. Sometimes light can be undesirable: soldiers in the field often have to forego the warmth

The Personality of Fire

of a fire for fear of being detected by the enemy. Often the light of a fire is an unused by-product: the light inside your home furnace doesn't serve any purpose. Throughout human history, however, fire has served us by making light.

One measure of human progress has been our growing ability to make light. You could make a long list of things important to you that depend on artificial light.

Liquids that Burn

Actually, liquids don't burn. You might swear to the contrary, but appearances can be deceiving. When you light a kerosene lamp, it is not the kerosene that burns but rather the vapors that are being released.

A flame cannot occur until the temperature is high enough to cause the vapors to escape. Have you ever seen a flaming dessert being served? It is usually a coating of brandy that is set afire. If you touch a match to a spoonful of brandy at room temperature, however, it will not ignite. You must first heat it until the temperature reaches the *flash point* of alcohol, touch a match to it so that the vapors start burning, and only then pour it on the dessert.

The flash point of a flammable liquid is the temperature at which the liquid produces vapors that can ignite. Even gasoline won't burn if it is below flash-point temperature. Don't try to prove this, however, since that temperature is much lower than that of your kitchen freezer. Kerosene, by contrast, is relatively safe because its flash point is considerably higher than normal room temperature. Hold a lighted match to a saucer of kerosene and it will not burn.

This might cause you to wonder how you can light a kerosene lantern with a match. The tip of the wick holds only a tiny bit of fuel, which is easily raised to flash temperature. The flame will not travel down the wick and ignite the fuel reservoir, however, because the temperature there is too low.

Even if a flammable liquid reaches its flash point, it will not burn unless it is ignited. This introduces a second way in which flammable liquids are classified: by their ignition temperature, or the degree of heat needed to produce a combustion in the fluid. The ignition temperature of gasoline is less than 300° C. Since the temperature of a burning cigarette is more than twice that, you can see why you shouldn't smoke while having your car filled with gas.

Still, you sometimes see motorists doing this very thing without mishap. What is the explanation? This time it is the flammability limit of gasoline. In order for gasoline fumes to explode, they must form a 1.4- to 7.6-percent mixture with air. If the percentage is beneath the lower limit, the mixture is too "lean," or thin, to burn. If it rises above the upper limit, the mixture is too rich to ignite.

Have you ever wondered how acetylene torches can burn underwater? The answer lies in the unusual flammability range of acetylene gas, from 2.5 percent all the way to 80 percent. The high upper

The Personality of Fire

limit means that there is enough oxygen even in water to enable combustion.

Putting out a Fire

To extinguish a fire, all you have to do is recall the fire triangle and remove one of the three essentials, fuel, oxygen, or heat. The most common method is to reduce the heat by hosing water onto the flames. Often, blankets of foam or chemicals are used to cut off the oxygen supply. When fighting forest fires, rangers plow a firebreak to deprive the fire of fuel.

Putting out liquid fires is a problem because of the *specific gravity* of most flammable liquids. This is the ratio of the weight of the liquid to the weight of the same volume of water. If the specific gravity ratio is less than 1, it means that the fluid is lighter than water and therefore will float on the surface and continue to burn. This is why fire fighters never use water on burning liquid fuels, as the water would cause the flames to spread.

If the flaming liquid is heavier than water, meaning that its specific gravity is greater than 1, it can be extinguished by floating water gently onto the surface of the liquid. Fire fighters call this *blanketing*.

Problems also attend gas fires because fumes also have varying densities. The density of a gas is the ratio of the weight of the gas to the weight of a similar volume of air. Ethyl ether has a density of 2.6, meaning that it is almost three times heavier than air. For this reason it will collect on the floor, and this

FIRE! ITS MANY FACES AND MOODS

explains why electrical outlets in hospitals are often located high on the walls to prevent a spark from igniting any ether that might escape and collect on the floor.

Heavy gases flow along the ground like an invisible river. Way back in 1915 this characteristic was the cause of a terrible flash fire in the town of Ardmore, Oklahoma. A gasoline truck was parked on a hot summer day. The heat was so great that the gasoline expanded and fumes began to escape. Unseen, they spread outward along the ground in a large circle. Something, probably a tiny spark, finally provided ignition, and flames flashed like lightning along the ground. Every building within a radius of 400 feet was destroyed. People walking blocks away had their clothing set afire by the sudden flash.

If your family uses a gas range for cooking, chances are that you have smelled leaking gas from time to time. This is because the density of cooking gas is about the same as that of the air, causing it to mix quickly in every direction. Actually, this is fortunate because it means you can smell the gas before it builds up to dangerous proportions. In a closed area, however, an undetected leak can create a bomblike threat. If you ever smell gas as you open the door to your house, do not go inside. Instead, leave the door open, and open additional doors and windows from the outside if possible. This will thin the gas-air mixture and in time it will be safe to enter and look for the leak. Never try to fix a gas leak yourself. Call your

The Personality of Fire

utility company for emergency service. Remember that leaking gas is poisonous as well as explosive.

Probably the biggest fire-fighting challenge is an oil well fire. A daredevil named Red Adair has become wealthy because of his courage and skill in putting them out. How does he do it? Strangely enough, with explosives. It's like blowing out a giant match with a superhuman puff of wind.

One of the most convenient sources of fire is the match. The next chapter will tell you all about this simple and useful invention.

13.

THE MIGHTY MATCH

How often have you heard someone say, "Give me a light."? The request is so common that you hardly give it a thought. Instinctively you strike a match. If the greatest of your great-grandfathers could have done the same thing he would have been worshipped as a god.

Humans have been walking the earth for half a million years. The land we now live on was populated first about 25,000 years ago when tribes trekked across a narrow bridge of land from Asia to Alaska where the Bering Strait now is. But it wasn't until the 1830s that the match was invented. If the span of the earth's existence were squeezed into a single year, it would mean that humans showed up a little after 11 P.M. on December 31, and that the match was invented almost at the stroke of the new year.

For a long, long time humans never even conceived of the idea of making fire. Fire was something to be

The Mighty Match

feared, something to run away from, as it is for wild animals. Anthropologists, scientists who study the evolution of the human race, can only guess how our ancestors learned that fire could serve them. Perhaps while running from a fire during winter months they noticed that the flames, though threatening, offered a welcome warmth. Maybe after such a fire people discovered that burned animals were easier to eat and that they tasted better.

However it happened, primitive humans learned that fire could be a friend as well as an enemy. Thereafter, tribes that roamed the continents always appointed one of their number to be the keeper of the fire. Since they thought that nature alone could kindle a flame, it was vital to keep one alive once it had been captured—to let it go out would be deserving of death. Each time the tribe stopped to make camp, they used the flame to light a cooking fire, and before they slept they set other fires to frighten away animal predators.

In America the Osage Indians kept a fire alive in a clump of tinder and lichen scraped from the inside of a hollow tree. The glowing ball was clamped inside the halves of a shell, and even after several days it would still be glowing when the shell was opened. On the other side of the world, islanders of New Guinea discovered the same trick, filling coconut shells with the glowing shavings of dried nuts.

Keeping a fire alive was a challenge from the era of the cave dwellers all the way into the Middle Ages.

FIRE! ITS MANY FACES AND MOODS

French householders disliked waking up on a cold morning to discover that the hearth fire had gone out during the night. When this happened it was a long, uncomfortable time before a new fire could be kindled. To avoid this they devised a simple procedure: before retiring for the night, they would gather the glowing coals into a heap and then cover them with a cone-shaped shield of brass or copper. In the morning when the shield was lifted, the flames revived immediately. The cone was called a *curfew,* derived from a French phrase meaning "cover the fire." The present-day meaning of the word comes from this original usage. A curfew now is an ordinance stating that citizens must retire from the streets by a certain time of the night.

In the course of time primitive humans learned that they didn't have to wait for a stroke of lightning to get fire. They discovered that there were ways of making it themselves. One of the most obvious ways was by the use of friction. We all know that we can warm our hands by rubbing them together. Why not use the same principle to start a fire? Independent of each other, people in different parts of the world learned that by twisting the point of a stick against a piece of dry wood a fire could be generated. The speed of the drilling was increased by looping the stick in a bowstring and moving the bow back and forth like a saw. This is the method used by boy scouts on camping trips today.

People also came to realize that it wasn't necessary

The Mighty Match

to travel to the sun to capture fire, as it was related in the myths. They could get fire from the sun without ever lifting a foot from the earth. As a youngster, you yourself probably discovered the magic of the magnifying glass. By focusing it correctly you can make the rays of the sun passing through the glass converge on a single point. This concentration of rays creates considerable heat, enough to start a fire.

Inca priests of ancient Peru boasted of great power because of their ability to make fire in a similar way. They held a golden bowl in such a way that the sun's rays reflected to a single point inside the bowl. A dry piece of linen or a wood chip held at this point would soon smoulder, giving the illusion of supernatural ability on the part of the priests.

A device very similar to the magic bowl is being used for outdoor barbecues, as well as for cooking in regions where there is no electricity. The bowl is much larger and it isn't made of gold, but the principle it employs is the same. It is called a solar oven. A large bowl-shaped metal shield is placed so that it collects sunrays and aims them at a single point where the food is placed. You won't have any luck after sunset, but otherwise the oven cooks just fine. Could you say that this is cooking without a fire? Not really. It's just that the fire is 200,000 miles away.

A third way of making fire was probably hit upon by accident, as are many discoveries. When certain kinds of rocks are hit together they make a spark. If enough of these sparks can be made to land on tiny

FIRE! ITS MANY FACES AND MOODS

chips of dry wood, a fire will result. It takes a long time—but your ancestors didn't have clocks to watch.

In colonial America the technique was essentially the same. By then, however, it had been found that flint and steel produce the most effective sparks, and that tinder was easier to light than wooden chips. Tinder was made from cotton or linen cloth, or from the powdered bark of certain trees. It was dried in an oven and removed just an instant before it was ready to break into flames. Then it was put into a tight tinderbox to be kept perfectly dry. A hunter or trapper in the wilderness considered his flint, steel, and tinderbox to be just as important as his knife and musket.

In 1780 a group of French chemists invented the *phosphoric candle.* Phosphorus is an unusual chemical that will burn just by being exposed to the oxygen in the air. To keep it from burning it has to be kept underwater or in a vacuum. The French inventors dipped a piece of paper in this chemical and then put it in an airtight glass tube. To start a fire, the tube was broken and the strip of paper would ignite. Awkward by today's standards, it was nonetheless the first step in the evolution of the mighty match.

An early version of the match was invented in 1827 by John Walker, an English pharmacist. Called *congreves,* they were wooden splints about 3 inches long, and tipped with a mixture of antimony sulfide, chlorate of potash, gum arabic, and starch. To light one

The Mighty Match

you drew the tip through a fold of glass paper, similar to sandpaper. It must have been an exciting experience, almost a Fourth of July sparkler. Sparks shot in all directions. More often than not, the congreves actually produced a flame.

A match which could be lit by striking it on any hard surface was invented in the 1830s by a French chemist, Charles Sauria. Phosphorus was the main ingredient used in the tip. At about the same time, a similar match was patented in the United States by Alonzo D. Phillips of Springfield, Massachusetts. Not long after production began on these matches, however, factory workers began to get sick and die. Their disease was diagnosed as necrosis of the jaw, and doctors determined that it was caused by the breathing of phosphorus fumes. Because matches were becoming so popular, they soon posed a serious threat to national health. Officials began to talk about a law to ban the manufacture of matches.

In 1900 the Diamond Match Company, still in business today, bought a French patent for a match that was tipped with a nonpoisonous phosphoric compound. When it was tested in the United States, however, it didn't work because of the climate difference. It appeared that the match industry was on the brink of ruin. Rescue came in 1911 when William A. Fairburn, a young naval architect, modified the French formula so that it would work over here.

In addition to providing a wonderful convenience,

FIRE! ITS MANY FACES AND MOODS

matches also created serious hazards. Accidental fires resulted because matches were so easy to ignite. A nibbling mouse could burn down a house, and clothing often caught on fire as a result of matches rubbing together inside a pocket. True to the old adage, necessity became the mother of invention. The safety match was designed in 1844 by Gustave E. Pasch, a Swedish chemist.

The idea was both simple and clever. Part of the chemicals needed to cause a flame were put into the tip of the match, and the remainder in the striking surface was glued to the side of the matchbox. The match tip would light only if it were rubbed against that surface. The introduction of the safety match practically eliminated accidental fires caused by matches themselves.

The first book-matches appeared in 1892 and were the invention of Joshua Pusey of Philadelphia. The original product, however, had the striking surface too close to the match tips for safety. Later, the Diamond Match Company purchased Pusey's patent and redesigned the cover. The striking surface was still on the side of the book that opened, however, giving many smokers a burned thumb every time a spark jumped and lit up the entire book. Today the striking surface has to be placed on the back of the matchbook. The growing popularity of cigarette smoking helped make book-matches a million-dollar industry.

When matches are hard to light it is usually because

The Mighty Match

of moisture. The tips of safety and book-matches are coated with a thin film of paraffin to keep them dry. During World War II a truly waterproof match was designed to meet the needs of men in combat conditions. Coated with a special chemical, the matches could be lit even after being underwater for several hours.

Because of the convenience of electric ranges and pilot-light gas stoves and ovens, very few matches are used in modern kitchens, but we keep them in reach for a multitude of other household purposes. And even though cigarette lighters have returned countless smokers to the era of flint and steel, the familiar request, "Have you got a light?" will always make us think of the mighty match.

As has been stressed throughout this book, fire should be treated with respect and care. Burns are some of the most painful injuries that you can have, and this is the topic of the next chapter.

14.
THE AGONY OF FIRE

Is there such a place as hell? If so, what do you think it's like?

Answers to these questions vary greatly, and no attempt will be made here to take a position. But one thing is certain. If in an afterlife there does exist a place of bodily torture, it would have to be a pit of flames. Few forms of human injury can exceed the agony of being burned.

Two-and-a-half million Americans suffer this pain each year, nearly a hundred thousand of them seriously enough to require hospitalization. Recuperation often takes years, and many victims never resume normal lives. Burn treatment can cost as much as $100,000.

Making these statistics even more gloomy is the fact that the figure is increasing, presenting a growing challenge to the medical profession; yet there are fewer than two hundred hospitals in the country that specialize in burn treatment.

The Agony of Fire

The skin is one of the most complex parts of the human body. An average adult body has more than 15 square feet of skin surface, and every square inch includes 3 feet of blood vessels, 12 feet of nerves, dozens of sweat glands, and millions of cells active in the constant process of dying and being replaced. Like the shell of a tortoise, your skin is the shield that protects you from external bacterial attack and maintains the correct body temperature. Any damage to your skin is a threat to your life, and the most deadly of all assaults is that by fire.

A *first-degree burn,* such as sunburn, is one that merely reddens the skin, and it constitutes no severe threat to your life. In a *second-degree burn* the outer layers of the skin are destroyed, and even after recovery there is usually permanent scarring. A *third-degree burn* is one in which all the layers of the skin are destroyed. The American Burn Association says that anyone with second- or third-degree burns on more than 20 percent of the body is considered to be seriously burned. The percentage is even lower for children and for the elderly.

A burn affects more than just the surface of the body—almost every organ is influenced. The greatest threat to life, however, comes from the loss of body fluids and the danger of infection from the breeding of bacteria in the region no longer protected by a layer of skin.

When you burn a finger on a hot stove, perhaps you

FIRE! ITS MANY FACES AND MOODS

have been taught to rub butter into the burn. For centuries essentially the same procedure was followed in hospitals, using more sophisticated forms of grease, of course. For years very little research was conducted to develop better treatments. Until 1960 there was only one hospital in the country equipped to handle severe burn cases effectively. It was the Brooke Army Medical Center in San Antonio, Texas. Today there are thirteen elaborate burn centers, but they hardly fill the total need.

What is done for a victim of severe burns when he or she is brought to a burn center? First, the person is bathed gently in a large tub. Sometimes incisions have to be made to reduce the pressure caused by swelling. While the patient is still in the tub it might be necessary to cut away loose skin. This operation causes unbelievable pain even when a powerful painkiller such as morphine is administered. Nonetheless, the pain is an encouraging sign because it means that the nerves are still functional.

If the patient loses too much body fluid, he or she will go into shock and probably die. To prevent this, large amounts of fluid are given intravenously, directly into a vein from a bottle suspended above the patient.

When the patient is finally placed in a bed, he or she is encased in a large thermal shield to prevent the loss of body heat. At the same time the burned areas are coated with medications that are infinitely more effective than the grease of bygone years.

For victims of third-degree burns, the most painful

The Agony of Fire

part of the treatment is the removal of dead skin. The process involves great loss of blood, and only small areas can be done at a time.

The final stage of treatment involves the grafting of new skin over the regions in which the skin has been lost. Just as hospitals have blood banks, burn centers have skin banks. Animal skin has been tried, particularly that of the pig, which resembles most closely the skin of humans. The results were negative because the body rejects foreign matter. The best method of skin grafting is to use the patient's own skin, taken from an unburned portion of the body.

It is now possible to make a little skin go a long way. Patches of skin taken from a healthy part of the body are put through a special roller called the Tanner Mesher. It stretches the skin into a netlike material which is almost nine times the original area. These mesh patches are stitched onto the burn regions, and in a few weeks the mesh grows together to cover the area completely. Because of this remarkable procedure, the need for repeated skin grafting operations has been greatly reduced.

Research continues, and every day burn victims' chances for survival become better. The future will see miniature lasers replacing scalpels for the removal of dead skin. Acceptable artificial skin might someday become a reality. Pressure masks capable of enclosing the entire body to eliminate ugly scarring have already been developed. Hypnosis is being studied as

FIRE! ITS MANY FACES AND MOODS

a way of blocking pain from the patient's consciousness. Psychiatrists are learning new ways of treating the mental trauma associated with severe burns.

No matter how many advances are made, however, the agony of burns will always be counted as one of the worst forms of human suffering. Even the weight of a bed sheet resting on burned skin can be unbearable. And this agony is shared by the doctors and nurses who must knowingly inflict pain in order to achieve an ultimate recovery. The strain is so severe that nurses specializing in burn medicine rarely have the strength to last out a full year. But the goal of their suffering is noble. The day is coming when victims of severe burns will no longer be doomed either to death or to life as hopeless cripples.

In the next chapter you will learn what simple precautions can help you avoid a tragic and painful experience with fire.

15.

THE CONSTANT MENACE

Every year fire claims the lives of 12,000 Americans. Statistics from the National Fire Data Center reveal that males, blacks, children, and old people dominate the statistics on fire fatalities. Every death is a painful event, but the real tragedy is that more than one-half of all fire deaths are the result of carelessness. They could be prevented.

Unfortunately, most young people are turned off by the slightest suggestion of a lecture on fire safety. You have to decide for yourself whether it's worth a little thought to keep yourself out of the grim statistics for fire victims. This short chapter describes some of the more common ways that fire poses a threat to your life, and how you can protect yourself from it.

Electricity

When young Ben Franklin conducted his famous kite experiment, he was courting death. If lightning had struck the brass key, which is a high conductor of

FIRE! ITS MANY FACES AND MOODS

electricity, attached to the kite string, the current would have traveled down the string, through Ben's body, and into the ground, electrocuting him. That's the way electricity behaves. It always seeks the easiest way to get into the ground.

The two hazards of electricity are electric shock and fire. When controlled, both can be used for the good. Mental patients can be cured by shock therapy, and the electrocardiogram tells doctors if your heart is behaving properly. In your home you use electricity in a dozen beneficial ways. The trick is to make electricity serve you, protecting yourself from its potential to kill.

Electricity flows through wires just as water flows through a pipe. Its pressure is called voltage, and its rate of flow is called current. Some materials like copper are called conductors because electricity flows easily through them. Because of the easy flow there is little resistance, so the wires in your home should never become hot. If too much current is forced through a small wire, however, heat will result and this could cause a fire. Plug a heavy-duty drill into an ordinary house outlet, for example, and you're in trouble. Fortunately, you have fuses or circuit breakers in the basement, which are designed to halt the electric flow before the lines become overloaded. Normal home wiring accepts a flow of 15 amperes (ampere = the unit by which current is measured). People sometimes unknowingly invite the possibility of fire by using a heavier fuse or by tampering with the

The Constant Menace

fuse box to prevent the emergency cut-off of electric current.

Some materials, such as nickel and tungsten, slow down the flow of electricity, becoming red-hot in the process. These metals are used to make the coils in irons and toasters. The conductor in an electric bulb becomes white-hot, and this is what gives you light to read by.

An *insulator*, such as glass or wood, is a material through which electricity cannot flow. Electric wires are wrapped in insulating materials. You should check all the wiring in your home to make sure that this covering is not cracked or frayed. If it is, electricity could escape and cause a fire.

You've heard the expression "short circuit." What does it mean? It's what happens when electricity takes a shortcut inside an appliance, by-passing some portion of its normal route through the coils. This places a heavier load on the rest of the circuit, producing abnormal heat and the threat of fire.

You should know some basic facts about toasters, air-conditioners, and other appliances that are used in your home.

Voltage Rating

A typical reading would be 105–120 volts. Since house voltage is normally 120 volts, such an appliance can be used safely. If air-conditioners or other units which use a lot of current are added to your home, you will have to have an electrician rewire the house to handle the additional load.

Frequency

Every appliance has a frequency rating, such as "60 cycles." Don't worry about this rating unless you are using the unit someplace other than in a typical American dwelling.

Current Rating

Measured in amperes, this tells you the rate of electric flow. Many appliances require more flow than ordinary house wiring can carry. Also, if you plug too many appliances into a single outlet, the house current won't be adequate to handle the total demand, the fuse will blow, and everything will black out.

Wattage Rating

You've heard of a 100-watt bulb. Its power rating, the amount of light it gives, is measured in watts. A 50-watt bulb will give only half as much light as a 100-watt bulb. If you divide the wattage rating by the voltage rating, you will get the approximate current, in amperes, that flows through an appliance.

You should be aware of some effective defenses against the potential electrical hazards in your home. Never overload a wall outlet. If you use an extension cord, make sure that it is the same type and size as the appliance's cord. Check all electric cords for signs of

The Constant Menace

wear. Never put cords under a rug, over a door, or over nails. Never use indoor wires outdoors or where they are exposed to water. Don't use electrical appliances near tubs or sinks. Don't use water on electrical fires.

Flammable Liquids

Several flammable liquids serve useful purposes around the home, but they should be handled with great caution. Look around your house and you might find one or more of the following.

Gasoline

Never store gasoline inside the house or attempt to use it as a cleaning agent. If it is necessary to use gasoline for running a home-maintenance appliance such as a lawn mower, store the gasoline in small quantities in a cool place such as the garage. Use only approved containers that have pressure-relief valves: a can isn't safe just because the word GAS is painted on it.

Never use gasoline to start an outdoor grill. Every year this foolish practice results in deaths and serious injuries. Never store gasoline in the trunk of a car. A simple rear-end collision could cause an explosion. Remember that a gallon of gasoline has the explosive force of fourteen sticks of dynamite.

Benzine

This is commonly used as a dry-cleaning agent and

FIRE! ITS MANY FACES AND MOODS

as fuel for cigarette lighters. It should be purchased only in small amounts and kept in a cool place in a tightly-sealed container.

Denatured Alcohol

This is used by athletes for massages. It is not extremely dangerous, but it is flammable and should be stored safely and used with care.

Paints, Removers, and Thinners

Most paints used around the house today are water-base and do not pose a fire threat. Oil-based paints and the turpentine used to thin them, however, are hazardous because their fumes are both flammable and toxic. Walls freshly painted with oil paint can cause the rapid spread of a fire.

Kerosene

Before the days of electricity, every household stored kerosene to use in the lamps. Many homes today keep kerosene for emergency lighting or to create a cozy atmosphere.

Not nearly as dangerous as gasoline, kerosene nonetheless has to be treated with respect. In addition to the fire hazard, remember that kerosene lamps and heaters consume oxygen, and that sufficient ventilation is vital in rooms where they are in use.

Explosions

Most common substances burn slowly. Gasoline and

The Constant Menace

other hydrocarbons ignite and burn in an instantaneous flash. This flash is called an explosion. Many home fires result from the slow accumulation of explosive vapors that escape from flammable liquids that are stored improperly. Remember also that dust can explode. A sufficient concentration in the air of dust particles from flammable materials such as cloth or wood is incredibly hazardous. An explosion in a flour mill can be as devastating as one in a gas refinery. Safety comes only with good ventilation.

Space Heaters

The first portable room heaters appeared in the 1920s. By the mid-1940s they were extremely common. Most of today's space heaters use gas, oil, or electricity. Some electric heaters contain a fan to circulate the heat more effectively.

Home accidents and fires resulting from the careless use of space heaters are quite common. Clothing is easily set on fire by space heaters, gas heaters can explode, and electric heaters can deliver a lethal shock. If you have to use space heaters in your home, be sure that they are carefully maintained, and use them only in accordance with the manufacturer's instructions.

Kitchen Fires

The kitchen oven and range are among a home's most important appliances. They can also be the most hazardous. Here are just three of the many reasons why.

FIRE! ITS MANY FACES AND MOODS

1. Flammable vapors can be ignited by the often-forgotten pilot light of a gas range.
2. Electric burner coils can reach a temperature of 550° C. After the burner is turned off, the coils, though no longer red, remain hot for a long time and can still burn.
3. Flammable fabrics can be ignited just by being near a hot oven or range.

A person who is cooking should never wear loose-fitting sleeves. Children should be trained not to climb on the range to get at the cookie jar. Ranges and ovens should be kept clean and clear of accumulated grease. If a pan of food begins to burn, turn the burner off and cover the pan with a lid, sliding it on from the front in order to deflect the flames. Never carry the pan to the sink, and don't ever use water on a grease fire.

Keep a dry-powder or carbon-dioxide extinguisher in the kitchen. Store it away from the areas where fires are most likely to occur so you will have access to it in the event of a fire.

Fireplaces

If you use a fireplace in your home, here are some do's and don'ts that will make for its safe operation.

Do use natural wood logs in preference to other materials. If you use artificial logs, follow instructions very carefully. Do inspect the fireplace regularly, keeping it and the chimney clean and in good repair.

The Constant Menace

Do always use a screen in front of the fireplace. Do keep the damper open enough to remove the smoke and gases, but not enough to let the fire burn out of control. Do keep the area near the fireplace clear of flammables. Do make sure that the fire is out before the family goes to bed. Do keep children away from the fire.

Don't burn more than one artificial log at a time. Don't burn coal, charcoal, or plastics in the fireplace, because poisonous gases can blow back into the house. Don't ever use gasoline or other flammable liquids to start or liven up a fire. Don't ever leave the fire unattended.

Similar cautions relate to the use of outdoor grills.

Fire in Public Places

Your home isn't the only place where the threat of fire lurks. Every time you ride in the family car or use public transportation, you face a risk. Eating at a restaurant, taking in a movie at the local theater, or dancing at your favorite disco could all end in fire. You shouldn't become obsessed with the danger of fire, living in constant fear to the point where life becomes a sequence of unbearable anxieties. At the same time, however, don't go to the opposite extreme and ignore danger. Take the middle ground by using common sense.

For example, when riding in a bus, note where the safety door is located. If you're sitting next to a punch-out window, take a minute to read the instructions.

FIRE! ITS MANY FACES AND MOODS

At the theater, restaurants, and clubs, make a note of the location of the nearest exit. Dance halls, particularly discos with far-out lighting, are extremely hazardous. Often they are crowded beyond the safety limit. Flammable decorations might be blocking the exits. Don't hesitate to report safety violations to police or fire officials.

Remember too that fires can happen at school. Many kids and even some teachers don't take fire drills seriously. The attitude seems to be that it won't happen here. But it did happen at Lakeview School in Collingwood, Ohio where 175 students died. It happened at Our Lady of Angels school in Chicago where 95 youngsters died. It happened at the Cleveland School in Beulah, South Carolina where 77 perished.

It's impossible to list every menace of fire; furthermore, no such list is worthy of memorization. Rather, it is important to know that hazards exist everywhere, and that most of them can be neutralized by the exercise of normal prudence. The United States Consumer Product Safety Commission has a free booklet, *Fire in Your Life*, which they will mail to you upon request. If you'd like a copy, write to them at Washington, D.C. 20207.

What would you do if a fire broke out in your home? The following chapter gives you some pointers that could save your life.

16.
HOME INFERNO

It's fun to imagine pleasant things. Wouldn't it be great to win a million-dollar lottery? Wouldn't it be wonderful to break the bank at a gambling casino? For a moment, however, imagine something that is not pleasant—something ugly. Imagine that soon you are going to die in a fire. Fortunately, the chance of this happening is just about as remote as winning a million. Yet statistics show that it is possible. It happens every year to twelve thousand Americans, and like it or not, you could become one of that number. What would it be like, those final minutes of your life? The following account is based on the way it happens to most victims of house fires.

It is late at night and fire breaks out while you are sleeping soundly in the comfort of your own bed. Not too many people die in daytime fires. When you are awake you can usually smell the early traces of smoke in time to get out of the building. Daytime fires can often be put out in a matter of minutes. Nighttime is something else. The entire family is asleep, and the

FIRE! ITS MANY FACES AND MOODS

early warnings aren't noticed. The fire spreads and the house begins to fill slowly with toxic gases.

Most people who die in house fires aren't burned. Without even waking up they die in their beds of asphyxiation—lack of oxygen. Sometimes they awaken in time to stumble toward a window or door, but they can't make it because there is no oxygen to feed their lungs.

Perhaps you awaken before things get quite that bad. The room is beginning to fill with smoke, but you are alert enough to know what is happening. Instinctively you reach for the lamp on your bedtable, but when you turn the switch nothing happens. Fire in the basement has shorted out the electricity.

Fire and darkness combine to be very frightening. You stumble out of bed and rush for the door, but you fall face forward onto the floor. You always were a little careless about leaving things scattered around. Luckily you don't break any bones and you are able to reach the door. But the moment you pull it open you are struck full force with a blast of searing heat and sickening smoke. You are almost overcome, but you are young and your strength prevails. You slam the door to survive for the moment, but now you are in a state of panic. You stagger to the window. If you can open it you are saved. But it won't open. It's stuck. This is the last thing you think as you crumple to the floor and die.

No flames have touched you. When the oxygen content of the air goes below 17 percent you begin to lose

Home Inferno

coordination. Below 16 percent you become irrational. When the oxygen level dips below 6 percent you can live no more than a few minutes.

If the temperature climbs to 150° C you will lose consciousness even if no smoke is present. The heated air atacks your lungs, and this combined with the carbon monoxide and hydrogen cyanide released by most fires leaves you no chance.

It's not a pretty story, and of course everyone hopes it won't happen to them or their loved ones. The odds are in your favor. But every hour of the day at least one American dies and thirty-four are injured, many of the latter crippled for life. It's worth the little effort it takes to prevent it from happening to you.

Planning

If you and your family were a platoon of soldiers facing an enemy force, you would first study the strengths and weaknesses of your position. You would anticipate the enemy's method of attack and prepare a counterattack. This is just common sense, and the same common-sense planning is needed if you hope to survive a possible attack by fire.

The first step is to have a family conference. Sit down with the building plans of your house and plot the best escape route from each room. Determine alternate routes should the initial routes be blocked by flames. Decide who should call whom should fire break out while everyone is asleep. Pick a spot outside

FIRE! ITS MANY FACES AND MOODS

the house where you will assemble after getting out of the building safely.

Inspection

Having a plan of escape won't help if the route is blocked at the time of emergency. Check windows and make sure that they are easy to open. If they are not, have them repaired. What about the window screens? Can they be removed easily? Some casement-window screens are bolted to the window frames, preventing easy egress. Many screens on combination windows work by push-buttons that are hard to find.

What is outside your bedroom window? Are you close to the ground? Is there a garage roof or a safe ledge that you can climb onto? If there is a fire escape, is it in good repair?

How many ways are there to get out of the house? Are doors easy to open, and are passageways uncluttered? No two homes or apartment buildings are completely the same, so no single set of rules will cover every situation. It's up to you and your family to make a plan appropriate to your own circumstances.

Practice

No amount of planning and inspection will help your family meet a fire emergency unless you practice. From time to time you should have a home fire drill. This doesn't mean that you have to startle the entire neighborhood by leaping out windows or rushing into

Home Inferno

the street. Just as it is at school, every movement should be calm, purposeful, and silent.

Blindfolded, each member of the family should be able to get out of bed and walk through the house to the front or rear door. Also, while blindfolded you should be able to open your window to ventilate the room and possibly to escape.

How long can you hold your breath? Long enough to get from your bed, through the house, and out the front door? Try it!

Lifesaving Tips

1. Sleep with your bedroom door closed.
2. If you smell smoke, never open a door until you have tested it. If it is hot or if smoke is seeping in around it, leave it closed.
3. If you open a window, make sure first that the door is closed. Otherwise you might create a draft that could suck fire and smoke into your room.
4. If you are trapped in your room, try to seal the cracks around the closed door. Open the window at the top and bottom, breathing through the bottom and allowing smoke to escape through the top.
5. In getting out of the house, keep your head low. Smoke rises and the air will be clearer near the floor.

FIRE! ITS MANY FACES AND MOODS

6. Try not to break a window as a means of escape. Glass shards make lethal daggers. If you must break a window, do it with a piece of furniture. Use a broom or any other stick to clear the frame of sharp glass before climbing through.
7. If you are escaping from a second-floor window, don't crouch and jump from the window sill. The drop would be about 13 feet and you could break a leg. Instead, hang by your hands from the ledge and then drop to the ground. Also, it's easier to get out of the window by standing on a chair and stepping out, rather than by climbing out.
8. As soon as you are clear of the fire, sound an alarm. Never assume that someone else has done so.
9. Never go back into a burning building. No material possession is worth more than your life. If a person is trapped inside, leave the rescue to the fire fighters who are trained and equipped. Many people have died trying to rescue someone who already had escaped through a different exit.
10. If a person's clothing is on fire, stop him or her from running, which only fans the flames. Knock the person to the ground and roll him or her on the ground or in a blanket until the flames are snuffed out.
11. Never try to fight the fire yourself. Call the

Home Inferno

fire department immediately so that precious minutes will not be lost.

Most people hate to call the fire department for a minor blaze for fear of appearing foolish. Such an attitude, however, often allows small blazes to become big ones. There are, of course, minor household emergencies that call for an immediate response. Here are a few that are common, along with recommendations for appropriate action.

1. *Burning food in the oven:* Close the oven door and turn off the heat. The lack of oxygen will starve the fire. Open windows and vents to let the smoke out.
2. *Smoke from an electric appliance:* Pull the plug if you can. If not, pull the fuse or circuit breaker that controls that line. (Do you know where these are?) Never use water on any flames until you are certain that the electricity has been cut.
3. *Smoke from the television set: Don't* try to pull the plug. Stay clear of the set. The picture tube could explode, and there is very high voltage inside. Turn off the main switch. If smoke continues call the fire department. If the smoke clears, do not use the set again until it is checked by a service man.
4. *Stove-top pan fire:* Cover the pan with a lid or a plate. Do not use water. Do not try to carry the pan to the sink.

5. *Deep-fat fire:* If you can get close, cover the pot with a metal lid. Don't try to fight the fire. Call the fire department.

You might wonder if homes are safer today than they were years ago. Most people would guess that they are. After all, think of all that must have been learned over the years. This is true in many ways. In one sense, however, modern homes are more of a fire hazard than were homes at the turn of the century. The explanaation is contained in a single word—*plastics.*

In the years of your great-grandmother everything inside the house was made of wood, cloth, and other natural materials. Today, furniture, rugs, drapery, and even ceilings and walls are made of synthetic compounds. Most plastics burn silently and give off poisonous gases that often can't even be smelled. Houses are often sealed tight for heating and air conditioning efficiency, and these explosive gases collect and ignite, causing what fire fighters call a *flashover.* Other plastic materials burn with a smoke so thick that firemen wearing masks have to grope for victims.

A good example of the plastics hazard was provided by a fire at New York's Kennedy International Airport in 1970. A huge room contained row on row of plastic chairs. One of these caught fire and as the flames spread from seat to seat, a cloud of gas formed at the ceiling. When flames leaped high enough the cloud exploded, turning walls of glass into flying glass shrapnel and spreading flames in every direction. In-

Home Inferno

stead of burning, the plastic ceiling melted, dropping fiery rain to the floor below. Amazingly, no lives were lost. Three months later, however, 145 teenagers from the St. Laurent-du-Pont School in France were not so lucky. They were enjoying a dance, and the hall was sprayed with plastic to give the appearance of a cave. The plastic caught fire, described by a witness as being "like a red panther in a small cage."

Two breaths of the toxic gases released by burning plastics are enough to make you keel over. For this reason many fire departments are requiring breathing apparatus as standard equipment for fire fighters at all fires, even if no smoke is visible. Fire fighters are often poisoned without knowing it, collapsing only after returning to the firehouse.

Returning to the question of improved home safety, one product developed in recent years makes modern homes safer from the danger of fire: the *electronic smoke detector*. A smoke detector is inexpensive, and most homes can be protected with only a few easily-installed units.

There are two types: one operates on an optical or photoelectric principle, and the other uses an ionization chamber. Both types are activated when tiny smoke particles enter the unit. The correct placement of these units, in decreasing order of importance, is as follows:

1. Outside each sleeping area.

FIRE! ITS MANY FACES AND MOODS

    2. At the top of the basement steps.
    3. In the living room, family room, or study if these areas are more than 15 feet from the bedroom detectors.
    4. Inside the bedroom of any person that smokes.

Don't try to save money by buying fire alarm units that are not fully approved. Buy in a reputable store and you will get satisfactory equipment. Follow carefully the installation and testing instructions on the box.

Some fire alarms are set off by heat rather than by smoke. For ordinary house conditions, however, these are not as good. You may be dead from smoke inhalation before enough heat builds up to activate the alarm.

A final important defense against a potential home inferno is the fire extinguisher. Before you buy one, however, you should know about the three classes of fires.

A fire fueled by ordinary materials such as wood, paper, or cloth is a *Class-A* fire. *Class-B* fires are fed by flammable liquids such as gasoline or kitchen fats. *Class-C* fires involve electrical equipment. All fire extinguishers must be labeled to indicate the kind of fire they are designed to extinguish. Many today are multipurpose.

Numerical ratings are also assigned to different extinguishers. The higher the number the more effective the unit will be in fighting larger fires. An extinguisher

Home Inferno

might be rated 1-A : 10-B,C. This would mean that the device will work on minor Class-A fires and would be effective against larger Class-B and Class-C fires. C-rated extinguishers always have dousing agents other than water.

For little money you can get a copy of *People and Fire*, a booklet published by the U.S. Department of Housing and Urban Development. Write to the Superintendent of Documents, U.S. Government Printing Office, Washington, D.C. 20402.

EPILOGUE
MANY FACES AND MOODS

As you read through the previous chapters and thought of the agony that fire can cause, perhaps you were tempted to wish that there would be no such thing as fire. Pause for a moment, however, and consider what the world would be like without fire.

It would take a long, long time to list the countless ways in which you depend directly or indirectly on fire. Without fire there would be no automobiles, airplanes, or steamships. Even your bicycle wouldn't exist because its steel frame and vulcanized tires are the offspring of fiery furnaces. There would be no such thing as a hot shower, a warm snow-covered house, or a patio cookout. Like the animals of the jungle, you would have to eat raw flesh and uncooked grain.

Your house would be a tent made of animal skins, a hut made from sun-hardened mud, or bundles of thatch because bricks come from a kiln. Without a blast-furnace there would be no nails, hammers, or saws. The windows that let in light and keep out cold wouldn't be possible, because glass was discovered accidentally in the glowing embers of a spent fire.

To envision yourself in a world without fire you

FIRE! ITS MANY FACES AND MOODS

would have to picture the world more than 50,000 years ago—the time when fire was yet to serve humans. Your tools and weapons would be shells and stones. Even the oldest of primitive tricks, the method of sharpening and hardening the tip of a wooden spear by burning and scraping its tip, would be taken from you.

Fire has been a source of wonder from the dawn of history to the present day. As you have read, its awesome force can inflict destruction, agony, and death. Yet, without it the human race would find it hard to survive. At the same time it is master and servant, friend and foe, devil and god. As the pendulum swings equally in opposite directions, so too does fire bestow both blessings and curses.

Will fire play the same role in the future as it has in the past? There is no doubt that it will do so for many centuries to come. Citizens in the far-distant future, however, will know scientific and technological advances that will dwarf the reality of the present.

The world's fuel supply is vast, but at the same time it is finite. If the human race doesn't first destroy itself with nuclear weapons, the age will come when natural deposits of coal, oil, and gas will be used up. What will happen then?

Scientists are already searching for answers to that question. The current energy crisis has set the stage for intensive research. Battery-driven cars will be a reality in the very near future. Solar energy is being harnessed more and more for heat and power. Despite

Many Faces and Moods

the fears of many, there is no way that nuclear power can be blocked from playing a vital role in the centuries to come.

If citizens of the ancient world could have looked at our world today, they would indeed be amazed. We too would be astonished if we could look into the world of the future. There are forces of nature still to be discovered. The gas, coal, and oil that are so vital to our present age may some day become as insignificant as the world's once crucial supply of whale oil.

Even if in some distant era fire were to lose its high rank among the elements of nature, it will never completely go away. The descendents of your great-grandchildren will probably still sing songs at campfires, and chances are that there will still be candles burning on their birthday cakes.

GLOSSARY

ALIDADE An instrument fitted with a telescope used for determining direction. Used on forest watchtowers to report the location of a smoke-rise.

ASPHYXIATION To become unconscious or die for want of oxygen. The most common cause of death attending fires.

AUTO-DA-FE Practiced during the period of the Inquisition, the burning at the stake of a person convicted of heresy.

BACKFIRING Setting a new fire in the direction of an existing fire in order to deprive the oncoming fire of its fuel supply.

BRITISH THERMAL UNIT The amount of heat needed to raise the temperature of one pound of water one degree Fahrenheit.

BUCKET BRIGADE A double line of volunteers passing buckets of water to a fire from a source of water, and back again for refilling.

CALORIE The amount of heat needed to raise the temperature of one gram of water one degree Celsius.

CELSIUS TEMPERATURE SCALE The thermometer scale on which the freezing point of water is

Glossary

0 degrees, and the boiling point is 100 degrees. Sometimes called the *centigrade* scale.

CONDUCTION   The process by which heat is transferred through physical contact.

CONVECTION   The process in which heat is transferred by rising upwards in currents.

FAHRENHEIT TEMPERATURE SCALE   The thermometer scale on which the freezing point of water is 32 degrees and the boiling point is 212 degrees.

FIRE   Rapid oxidation that results in light, flame, and heat. (See OXIDATION)

FIRE LINE   A path cleared by fire fighters to halt the spread of a fire.

FIRE MARK   No longer in use, an emblem attached to the face of a building identifying the company that insures the building. A fire brigade would put out a fire only if the fire mark belonged to the company that employed them.

FIRE MARSHAL   A fire department detective who investigates cases of suspected arson.

FIRE TRIANGLE   The three essential ingredients of a fire: fuel, heat, and oxygen.

FLAME   The glowing gaseous part of a fire. (See FIRE)

FLAMMABILITY LIMIT   The range of percentage mixture with air in which flammable vapors will ignite. Too lean or too rich a mixture makes ignition impossible.

FLASH POINT   The temperature at which a flammable liquid gives off vapors that can ignite.

179

FIRE! ITS MANY FACES AND MOODS

GEOTHERMAL POWER Power derived from the heat which is stored beneath the earth's surface.

GREEK FIRE A highly flammable mixture of naphtha and pitch used as a weapon in seventh century warfare.

HEAT The energy that is given off when rapid molecular activity is generated in a substance.

HEAT OF COMBUSTION The amount of heat released that varies according to the nature of a burning substance. It is measured in *British Thermal Units* or in *Calories*.

HYDROCARBON An organic compound containing only carbon and hydrogen atoms. Highly flammable, the hydrocarbon is found primarily in petroleum and natural gas.

IGNITION TEMPERATURE The heat needed to produce a self-sustaining flame in a flammable liquid.

KINDLING POINT The temperature at which a substance will catch fire.

MAGMA Vast pools of molten rock produced by the heat of pressure beneath the surface of the earth.

MOLOTOV COCKTAIL Used by the Russians during World War II, a bottle filled with gasoline and hurled as a weapon against enemy tanks and bunkers.

NAPALM Developed by the United States in World War II, a jellied gasoline used in bombs and sprayed from flamethrowers.

Glossary

OXIDATION The process in which the surface atoms of some substances combine with oxygen atoms in the surrounding air.
PHLOGISTON Once thought to be the hidden ingredient that explained why some things could burn.
PHOSPHORIC CANDLE The first step in the evolution of the match, a phosphorus-tipped slip of paper that ignited when removed from a glass tube, exposing it to the air.
PITCH A tar-like substance capable of burning for a period of time, used to coat the tips of torches during the Middle Ages.
PYROMANIAC A mentally unbalanced person with an irresistable impulse to start fires.
PYROTECHNICS The art of firework display.
RADIATION The transfer of heat by means of invisible waves that spread in every direction from a source of heat.
RATTLE WATCH A night patrol used in colonial times to alert the town in case of a fire. Loud wooden rattles were used to sound the alarm.
SCORCHED EARTH A defensive tactic in which a retreating army burns anything that might be of use to the advancing enemy.
SHORT CIRCUIT A fire hazard in electrical appliances that results when a current bypasses part of its normal channel, thus causing the remainder of the circuit to overheat.
SMOKE Unburned molecules, usually carbon, which

FIRE! ITS MANY FACES AND MOODS

flow from a fire as a result of incomplete combustion.

SMOKE DETECTOR An instrument designed to sound an alarm when it detects the presence of smoke molecules, and now used extensively in homes and apartments.

SMOKE JUMPERS Fire-fighting parachutists dropped from airplanes into strategic locations for combatting the flames.

SPONTANEOUS COMBUSTION Self-ignition of a substance that results from heat produced by internal chemical action.

TEMPERATURE The measure of how fast the molecules of a substance are vibrating. (See CELSIUS and FAHRENHEIT)

THERMAL EXPANSION The property of many substances whereby they expand when heated and contract when cooled.

TINDER BOX Carried in one's pocket in the days before matches were invented, it contained a dry flammable powder that was easy to ignite with a spark from flint and steel.

TORCH Gangster term for a professional arsonist.

TRAILER A string of flammable material laid out by an arsonist and soaked with a flammable liquid to hasten the spread of a fire.

VESTAL VIRGINS Young girls dedicated to keeping the sacred fire alive in the ancient Roman temple of Vesta, the goddess of fire.

SUGGESTED FURTHER READINGS

Butler, Hal. *Inferno.* Chicago: Henry Regnery Company, 1975.
Clevely, Hugh. *Famous Fires.* New York: John Day Company, 1958.
MacDonald, Gordon A. and Hubbard, Douglas H. *Volcanoes of the National Parks of Hawaii.* (Booklet) Honolulu: Tongg Publishing Company, Inc., 1974.
Morris, John V. *Fires and Firefighters.* Boston: Little, Brown and Company, 1955.
Owen, Howard R. *Fire and You.* New York: Doubleday and Company, Inc., 1977.
Rittmann, A. and L. *Volcanoes.* New York: G. P. Putman's Sons, 1976.
Smith, Dennis. *Report from Engine Co. 82.* New York: Saturday Review Press, 1972.
Stewart, George Rippey. *Fire* (a novel). New York: Random House, 1948.

INDEX

Acrophobia, 81, 118
Adair, Red, 139
Aerial ladder, 98, 106
Aerial platform, 107
Africa, 68, 69, 72
Alarm. *See* fire alarm
Alaska, 80, 140
Alcohol, 130, 135
America, 31, 48, 87, 97, 130, 141, 144. *See also* United States
American Burn Association, 149
American Indian, 43, 48, 85-86
American Insurance Association, 92
American Revolution, 46
Antiaircraft shells, 52
Apollo spacecraft, 66
Appliances, 155, 156, 157, 159, 169
Archimedes, 43
Arson, 86, 87-95
Arson task force, 88, 93, 95
Artillery shell, 51-52
Astronauts, 65-66
Atoms, 17, 121, 122, 123, 124

Auto-da-fé, 21-22

Backfiring, 83-84
Bacon, Roger, 48-49
Benzine, 157-158
Beverly Hills Supper Club, 63
Blackout, 87-88
Breathing apparatus, 115, 116, 171
British Thermal Units (BTUs), 127
Brooke Army Medical Center, 150
Bucket Brigade, 97, 98, 99
Burns, 30, 112, 147, 148-152

Caesar, Julius, 13, 21, 76
Calixtus III, Pope, 22
Calories (cal.), 127
Candle, 14, 18, 20, 31, 32, 41, 49, 132, 144, 177
Cannon, 48, 49-50, 51
Carbon, 123, 124, 134
Carbon monoxide, 134, 165
Carter, President Jimmy, 77-78
Celsius, Anders, 131

Index

Celsius scale, 130-131
Cerro Rico, 77
Chabert, Julian Xavier, 27-28
Charles VII of France, 50
Chemist, 30, 144, 145, 146
Chicago Fire. *See* Great Chicago Fire
Chinese New Year, 49
Cinder cone, 70
Civil Service examination, 117-118
Civil War, 47, 52, 101
Cleveland Hospital fire, 65
Coal, 22, 29, 38, 39, 122, 127, 142, 161, 176, 177
Coast Guard, 63
Coconut Grove fire, 63-64
Collingwood fire, 65, 162
Colorado River, 37
Colosseum, 49
Columbus, 21
Conduction, 132, 133
Congreves, 144-145
Constantinople, 48, 50
Convection, 133, 134
Convection oven, 40, 41
Curfew, 142
Current rating, 156

Dark Ages, 103
Denatured alcohol, 158
Dial thermometers, 132
Diamond Match Company, 145, 146

Easter Sunday, 19, 20
Electricity, 33, 37, 38, 41-42, 78, 153-155, 158, 159, 164, 169
Empire State Building, 108-109
Energy, 77-78, 176
England, 31, 51, 100
Epimetheus, 16
Ethyl ether, 137-138
Eunus, 25
Explosives, 51, 158-159

Fahrenheit, Gabriel D., 130
Fahrenheit scale, 130, 131
Fairburn, William A., 145
FBI, 91
"Fire," 86
Fire alarm, 58, 97, 101, 107, 109-110, 112, 114, 118, 168, 172
Fireboat, 106, 108
Fire chief, 111, 114
Fire commissioner, 114
Fire company. *See* fire department
Firecrackers, 49, 51
Fire department, 95, 98, 99, 100, 103, 108, 110, 112, 113, 114, 117, 118, 169, 170, 171
Fire-eating, 26
Fire engine, 36, 102, 104, 108, 111
Fire extinguisher, 58, 113, 160, 172-173

Index

Fire festival, 19-20
Fire fighters, 59, 96, 97, 98-99, 100, 101, 102, 104, 105, 107, 110, 111-119, 127, 137, 168, 170, 171
Firehorses, 101-102
Firehouse dogs, 102
Fire insurance, 91-92
"Fire in Your Life," 162
Fire line, 83, 84, 86
Fire mark, 100
Fire marshal, 93-95
Firemen. *See* fire fighters
Fire of London (1666), 103
Fireplace, 16, 24, 31, 160-161
Fire Prevention Week, 113
Fire-walking, 29-30
Fireworks, 49
Flamethrowers, 44, 45
Flammable fabric, 160
Flammable liquids, 135-136, 137, 157-159, 161
Flash point, 135-136
Flint, 52, 144, 147
Forest fire, 18, 55-56, 79, 80-86, 104, 137
Foy, Eddie, 58-59
France, 13, 22, 171
Franklin, Benjamin, 133, 153
Franklin Stove, 133
Frequency, 156
Fuel, 37, 39, 70, 122-123, 127, 134, 136, 137, 176
Furnace, 27, 28, 38-39, 134, 135, 175

Galilei, Galileo, 129
Gas, 38, 74, 76, 121, 134, 137-139, 159, 176, 177
Gas fire, 137-138
Gasoline, 27, 44, 94, 122, 123, 127, 134, 135, 136, 138, 157, 158-159, 172
Gas range, 138, 147, 160
Gaul, 13, 21
General Slocum fire, 61-62, 63
Geothermal power, 77
Geyser, 78
Great Chicago Fire of 1871, 54-55, 98, 113
Greece, 13, 24
Greek fire, 48
Greeks, 16, 17
Gunpowder, 48-49, 50, 51, 52

Haleakala National Park, 67
Hand pumper, 98, 99, 101
Harrison, President, 80
Hawaii, 67, 71, 78
Heart attacks, 112
Heat, 25, 28, 40, 121, 123, 124, 126-129, 130, 132-134, 135, 137, 138, 154, 155
Heaters, 158, 159
Heimaey, 70-71
Hell, 22-23
Helmet, 115-116
Hemingway, Ernest, 72
Heraclitus, 17-18
Herculaneum, 74
Hercules, 17

187

Index

Hero, 25, 34
Hindenberg, 74
Hoover Dam, 37
Hot-water heating system, 39-40
Houdin, Robert, 27
Hydroelectric plant, 37-38
Hydrogen, 121, 122, 123, 124
Hypnosis, 151-152

Iceland, 70-71
Incas, 77, 143
India, 48, 49
Indonesia, 69, 72
Industrial Revolution, 33
Inquisition, 21-22
Insurance Information Institute, 54
Iroquois Theater fire, 57-60
Ivan III, Duke, 46

Jet airplanes, 34, 37
Joan of Arc, 22
John F. Kennedy carrier, 87

Kennedy, President, 16
Kerosene, 65, 135-136, 158
Kilaueau, 67-68, 71, 78
Kilimanjaro, 72
Kitchen fires, 159-160
Korean War, 44
Krakatoa, 69, 72

Ladderman, 112-113

LaGuardia, Mayor Fiorello, 114
Lassen Peak, 71
Lichen, 141
Lifesaving tips, 167-170
Light, 32, 33, 41, 134-135, 158
Lipari Islands, 77
Liquid fire, 94, 137
Liquids, 30, 121, 122, 126, 129, 135-137
Liquified gas, 115
London Blitz, 45, 52

Magic, 19, 23, 24-32
Marriage, 14, 15
Martinique, 72
Massachusetts Firefighting Academy, 115
Match, 51, 135, 136, 139, 140, 144-147
Mauna Loa, 67
Mercury, 121, 130
Mercury thermometer, 128, 131-132
Metal, 18, 26, 27, 33, 41, 44, 155
Microwave oven, 40-41
Middle Ages, 18, 19, 20, 22, 26, 141
Molecules, 40, 121, 122, 123, 124, 128, 132, 134
Mont Pelée, 72
Moscow, 45-46, 47
Mount Katmai, 71
Mount Vesuvius, 72-73

Index

Musket, 48, 49, 52, 144
Mythology, 16

Napoleon, 47
Napalm, 44-45
Naphtha, 48, 94
National Fire Data Center, 153
Newcomen engine, 35-36
Newcomen, Thomas, 35
New York City, 46, 87, 93, 101, 104, 106, 108
New York Department of Fire Prevention, 65
New York's Triangle fire, 65
Nuclear energy, 42, 78, 123
Nuclear power, 42, 52-53, 177

Oil, 26, 38, 39, 159, 176, 177
"Oil of Medea," 48
Oil well fire, 139
Old Faithful, 78
Ore, 33, 77
Oxidation, 123
Oxygen, 30, 66, 121, 122, 123-125, 127, 134, 137, 144, 158, 164-165, 169

Pacific Plate, 71
Pagan rituals, 21
Paint remover, 158
Paint thinner, 158
Paramedical service, 113
Paschal Candle, 20
Pasch, Gustave E., 146

Pearl Harbor, 45
"People and Fire," 173
Peshtigo Forest fire, 55-56
Phillips, Alonzo D., 145
Philosophers, 17-18
Phlogistan, 120-121, 122
Phosphoric candle, 144
Phosphorus, 144, 145
Plastic, 115, 161, 170-171
Plate tectonics, 68-69, 75
Pompeii, 66, 69, 72-74
Porta, Giovanni della, 35
Priestly, Joseph, 122
Prometheus, 16, 17
Property Insurance Loss Register, 92
Public places, 161-162
Pumice, 76
Pumpers, 103-104, 105, 106, 113. *See also* hand pumpers
Pump man, 105-106
Punic Wars, 45
Pusey, Joshua, 146

Radiation, 133-134
Roman catapults, 45
Romans, 13-15, 45, 102, 103
Rome, 13, 14, 96, 109
Roof tanks, 104
Ross, British General, 46
Rubber, 33, 44

Salvage men, 113
San Carlos de la Rápata fire, 64

Index

Sauria, Charles, 145
Savery, Thomas, 35
Schwarz, Berthold, 49
Sea battles, 50-51
Sementini, 30
Sherman, General William T., 47
Sierra Nevadas, 86
Skin grafting, 151
Smoke, 19, 31, 48, 61, 62, 66, 94, 104, 111, 112, 113, 133, 134, 163, 164, 165, 167, 169, 171
Smoke detector, 171-172
Smoke inhalation, 112, 172
Soil, 75-76, 77
Solar oven, 143
Solids, 121, 127
South American, 68, 77
Spain, 21, 22, 51
Spanish Armada, 51
Spirits, 14-15, 32, 49
Spitcat, 86
S.S. High Flyer, 65
Stabiae, 74
Stahl, Georg, 120
Stanley Steamer, 36
Steam, 27, 28, 33-36, 37, 38, 121, 127
Steam engine, 33, 34, 35-36, 101, 102
Steam turbine, 37-38
Steel, 33, 52, 144, 147
Stewart, George R., 86
Stromboli, 71

Superstition, 19, 30-32
Suspended Maneuvering System (SMS), 104-105

Taft, President William, 63
Tambora, 72
Tanner Mesher, 151
Temperature, 94, 121, 127, 128-130, 131, 135, 136
Thermometer, 126, 129-130, 131-132
Tinder, 141, 144
Toxic gases, 73, 164, 170, 171
Turnout coat, 116

United States, 44, 67, 71, 77, 80, 95, 105, 111, 130, 145. *See also* America
United States Consumer Product Safety Commission, 162
U.S. Department of Housing and Urban Development, 173

Van Schaick, William, 61-62, 63
Vapor, 36, 127, 135, 158, 160
Verdun, battle of, 44
Vesta, 15
Vestal virgins, 15
Vesuvius, 74
Vietnam War, 44
Volcano, 66, 67-79
Voltage rating, 155-156
Volunteer fireman, 118

Index

Walker, John, 144
War of 1812, 46
Water cannon, 106, 113
Water damage, 113
Water tower, 107
Waterwheels, 37-38
Wattage rating, 156-157
Watt, James, 33-34, 35-36

World War I, 44
World War II, 43, 44, 52, 147

Yellowstone National Park, 78
Yellowstone Timber Reserve, 80

Zeus, 16-17

ABOUT THE AUTHOR

James O'Donnell has written and published several books aimed at the interests of young readers. The father of three children, he is on the faculty of Saint Cecilia High School, Englewood, New Jersey. He resides in Maywood, New Jersey.